T0332384

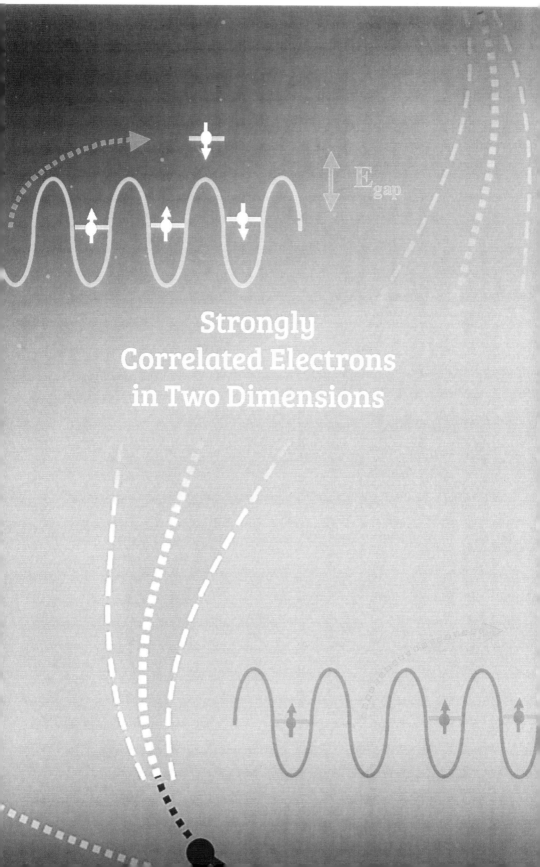

Strongly Correlated Electrons in Two Dimensions

Strongly Correlated Electrons in Two Dimensions

edited by

Sergey Kravchenko

PAN STANFORD PUBLISHING

Published by

Pan Stanford Publishing Pte. Ltd.
Penthouse Level, Suntec Tower 3
8 Temasek Boulevard
Singapore 038988

Email: editorial@panstanford.com
Web: www.panstanford.com

British Library Cataloguing-in-Publication Data
A catalogue record for this book is available from the British Library.

Strongly Correlated Electrons in Two Dimensions

ISBN 978-981-4745-37-6 (Hardcover)
ISBN 978-981-4745-38-3 (eBook)

Printed in the USA

Contents

Preface   ix

1 Wigner–Mott Quantum Criticality: From 2D-MIT to ^{3}He and Mott Organics   1
  V. Dobrosavljević and D. Tanasković
  1.1 MIT in the Strong Correlation Era: The Mystery and
       the Mystique   2
  1.2 Phenomenology of 2D-MIT in the Ultraclean Limit   4
       1.2.1 Finite-Temperature Transport   5
       1.2.2 Scaling Phenomenology and Its Interpretation   6
       1.2.3 Resistivity Maxima in the Metallic Phase   9
       1.2.4 Effect of Parallel Magnetic Fields   11
       1.2.5 Thermodynamic Response   13
  1.3 Comparison to Conventional Mott Systems   15
       1.3.1 Mott Transition in ^{3}He Monolayers on
              Graphite   16
       1.3.2 Mott Organics   18
  1.4 Theory of Interaction-Driven MITs   22
       1.4.1 The DMFT Approach   23
       1.4.2 The Mott Transition   25
       1.4.3 Correlated Metallic State   27
       1.4.4 Effective Mass Enhancement   28
       1.4.5 Resistivity Maxima   31
       1.4.6 Quantum Criticality and Scaling   34
              1.4.6.1 Is Wigner crystallization a Mott
                       transition in disguise?   37
  1.5 Conclusions   41

2 Metal–Insulator Transition in a Strongly Correlated Two-Dimensional Electron System   **47**

A. A. Shashkin and S. V. Kravchenko

2.1 Strongly and Weakly Interacting 2D Electron Systems   47
2.2 Zero-Field Metal–Insulator Transition   50
2.3 Possible Ferromagnetic Transition   53
2.4 Effective Mass or g-Factor?   56

3 Transport in a Two-Dimensional Disordered Electron Liquid with Isospin Degrees of Freedom   **65**

Igor S. Burmistrov

3.1 Introduction   65
3.2 Nonlinear σ-Model   74
   3.2.1 Nonlinear σ-Model Action   74
   3.2.2 Physical Observables   77
   3.2.3 One-Loop Renormalization   78
      3.2.3.1 Perturbation theory   78
      3.2.3.2 One-loop renormalization of physical observables   79
   3.2.4 One-Loop RG Equations   81
   3.2.5 Conductivity Corrections due to Small Symmetry-Breaking Terms   82
   3.2.6 Dephasing Time   83
3.3 Spin–Valley Interplay in a 2D Disordered Electron Liquid   84
   3.3.1 Introduction   84
   3.3.2 Microscopic Hamiltonian   84
   3.3.3 $SU(4)$ Symmetric Case   87
   3.3.4 $SU(2) \times SU(2)$ Case   88
   3.3.5 Completely Symmetry-Broken Case   90
   3.3.6 Discussion and Comparison with Experiments   93
3.4 2D Disordered Electron Liquid in the Double-Quantum-Well Heterostructure   98
   3.4.1 Introduction   98
   3.4.2 Microscopic Hamiltonian   98
      3.4.2.1 Estimates for interaction parameters   102
   3.4.3 One-Loop RG Equations   103
   3.4.4 Dephasing Time   105

3.4.5   Discussion and Comparison with Experiments   106
3.5   Conclusions   109
A.1   Appendix   109

4   Electron Transport Near the 2D Mott Transition   117
Tetsuya Furukawa and Kazushi Kanoda
4.1   Mott Transition   117
4.2   Theoretical Investigations of the Mott Transition   119
        4.2.1   Dynamical Mean Field Theory   119
        4.2.2   Quantum Criticality of the Mott Transition   121
4.3   Organic Materials: Model Systems of the Mott Physics   123
4.4   Temperature–Pressure Phase Diagram   125
4.5   Critical Phenomena around the Critical End Point   127
4.6   Quantum Criticality at Intermediate Temperatures   131
        4.6.1   Resistivity in the Crossover Region   131
        4.6.2   Scaling Analysis   134
        4.6.3   Comparison of the Experimental Results with
                   the DMFT Predictions   138
4.7   Summary   140

**5   Metal–Insulator Transition in Correlated Two-Dimensional
     Systems with Disorder   145**
Dragana Popović
5.1   2D Metal–Insulator Transition as a Quantum Phase
        Transition   145
5.2   Critical Behavior of Conductivity   149
        5.2.1   Role of Disorder   150
                   5.2.1.1   Low-disorder samples   150
                   5.2.1.2   Special disorder: local magnetic
                                   moments   154
                   5.2.1.3   High-disorder samples   157
        5.2.2   Effects of the Range of Electron–Electron
                   Interactions   161
        5.2.3   Effects of a Magnetic Field   165
        5.2.4   Possible Universality Classes of the 2D
                   Metal–Insulator Transition   167
        5.2.5   Metal–Insulator Transition in Novel 2D
                   Materials   169

5.3 Charge Dynamics Near the 2D Metal–Insulator
Transition and the Nature of the Insulating State   171
    5.3.1   High-Disorder 2D Electron Systems   173
    5.3.2   Low-Disorder 2D Electron Systems   177
5.4 Conductor–Insulator Transition and Charge Dynamics
in Quasi-2D Strongly Correlated Systems   178
5.5 Conclusions   182

**6 Microscopic Theory of a Strongly Correlated
Two-Dimensional Electron Gas**   **189**
M. V. Zverev and V. A. Khodel
6.1 Introduction   189
6.2 Ab initio Evaluation of the Ground-State Energy and
Response Function of a 2D Electron Gas   191
6.3 Ab initio Evaluation of Single-Particle Excitations of a
2D Electron Gas   196
6.4 Disappearance of de Haas–van Alphen and
Shubnikov–de Haas Magnetic Oscillations in MOSFETs
as the Precursor of a Topological Rearrangement of
the Landau State   203
6.5 Conclusion   215

Index   219

Preface

Properties of strongly correlated electrons confined in two dimensions are a forefront area of modern condensed matter physics. Two-dimensional (2D) electron systems can be realized on semiconductor surfaces (metal-insulator-semiconductor structures, heterostructures, quantum wells); other examples include electrons on a surface of liquid helium or a single layer of carbon atoms (graphene). In some of these systems, Coulomb repulsion between electrons is small compared to the kinetic energy of electrons; such systems can be well described by Fermi liquid theory introduced by Landau in 1956. However, when the energy associated with the Coulomb interactions becomes larger (sometimes by orders of magnitude or even more) than the Fermi energy, perturbation theories fail and one may expect novel states of matter to form.

In a zero magnetic field, idealized (noninteracting) 2D electrons were predicted by the "Gang of Four" (Abrahams, Anderson, Licciardello, and Ramakrishnan, 1979) to become localized in the limit of zero temperature, no matter how weak the disorder in the system. Weak interactions between electrons are expected to contribute to the localization (Altshuler, Aronov, and Lee, 1980). Therefore, it came as a surprise when the metallic (delocalized) state and the metal–insulator transition were observed in a 2D electron system formed in low-disordered silicon transistors (Kravchenko et al., 1994). Since then, a tremendous effort has been made, in both theory and experiment, to produce an adequate understanding of the situation; however, a consensus has still not been reached.

In the limit of very strong interactions, electrons are supposed to crystalize into a lattice to minimize their repulsion energy (Wigner, 1934). A classical Wigner crystal has indeed been realized for electrons on the surface of liquid helium. Although indications exist that Wigner crystallization also occurs in very dilute electron systems on semiconductor surfaces (where the crystal should be quantum), the "smoking-gun evidence" has never been obtained.

These are just two examples of many outstanding unsolved problems in the physics of strong correlations in two dimensions.

This book, intended for advanced graduate students and researches entering the field, contains six chapters. In Chapter 1, a review is given on the recent theoretical work exploring quantum criticality of Mott and Wigner–Mott transitions. The authors argue that the most puzzling features of the experiments find natural and physically transparent interpretations based on this perspective.

Chapter 2 is devoted to experiments on very clean and very dilute 2D electron systems. Experimental results on the metal–insulator transition and related phenomena in such systems are discussed. Special attention is given to recent results for the strongly enhanced spin susceptibility, effective mass, and thermopower in low-disordered silicon transistors.

In Chapter 3, the author shows how spin and isospin degrees of freedom affect low-temperature transport in strongly interacting disordered 2D electron systems and explains experimentally observed temperature and magnetic field dependencies of resistivity in silicon-based systems.

In Chapter 4, recent experimental studies on the Mott transitions of layered organic materials are reviewed with an emphasis on quantum-critical transport. The authors show that in the vicinity of the Mott transition, different kinds of phases emerge, such as antiferromagnetic Mott insulators, quantum spin liquids, Fermi liquids, and unconventional superconductors.

Chapter 5 is a review of experimental results obtained on 2D electron systems with different levels of disorder. In particular, the author shows that sufficiently strong disorder changes the nature of the metal–insulator transition. Comprehensive studies of the charge dynamics are also reviewed, describing evidence that the metal–insulator transition in a 2D electron system in silicon should be viewed as the melting of the Coulomb glass.

Finally, in Chapter 6, a microscopic theory of a strongly correlated 2D electron gas is presented. The authors suggest an explanation of the divergence of the effective electron mass experimentally observed in silicon-based 2D structures. Possible condensation of fermions in 2D electron systems, closely related to the condensation of bosons in superconductors or in superfluids, is also discussed.

I hope that this book will stimulate further developments in the physics of strongly correlated electrons in two dimensions and lead to many discoveries of yet unforeseen new physics.

Sergey Kravchenko

Chapter 1

Wigner–Mott Quantum Criticality: From 2D-MIT to ^{3}He and Mott Organics

V. Dobrosavljević[a] and D. Tanasković[b]

[a] *Department of Physics and National High Magnetic Field Laboratory, Florida State University, Tallahassee, Florida 32306, USA*
[b] *Scientific Computing Laboratory, Center for the Study of Complex Systems, Institute of Physics Belgrade, University of Belgrade, Pregrevica 118, 11080 Belgrade, Serbia*
vlad@magnet.fsu.edu, tanasko@ipb.ac.rs

Experiments performed over the last 20 years have revealed striking similarities between several two-dimensional (2D) fermion systems, including diluted two-dimensional electron liquids in semiconductors, ^{3}He monolayers, and layered organic charge transfer salts. These experimental results, together with recent theoretical advances, provide compelling evidence that strong electronic correlations—Wigner–Mott physics—dominate the universal features of the corresponding metal–insulator transitions (MITs). Here we review the recent theoretical work exploring quantum criticality of Mott and Wigner–Mott transitions and argue that most puzzling features of the experiments find natural and physically transparent interpretation based on this perspective.

Strongly Correlated Electrons in Two Dimensions
Edited by Sergey Kravchenko
Copyright © 2017 Pan Stanford Publishing Pte. Ltd.
ISBN 978-981-4745-37-6 (Hardcover), 978-981-4745-38-3 (eBook)
www.panstanford.com

1.1 MIT in the Strong Correlation Era: The Mystery and the Mystique

The key physical difference between metals and insulators is obvious, even at first glance. Its origin, however, has puzzled curious minds since the dark ages of Newton's alchemy (Dobbs, 2002); centuries-long efforts have failed to unravel the mystery. Some basic understanding emerged in the early 1900s with the advent of quantum mechanics, based on the nature of electronic spectra in periodic (crystalline) solids. The corresponding band theory of solids (BTS) (Ashcroft and Mermin, 1976) provided a reasonable description of many features of good metals, such as silver or gold, but also of good insulators, such as germanium and silicon.

From the perspective of the BTS, metals and insulators represent very different phases of matter. On the other hand, the rise of electronic technology, which rapidly accelerated in the second half of the 20th century, demanded the fabrication of materials with properties that can be conveniently tuned from the limit of good metals all the way to poor conductors and even insulators. This requirement is typically satisfied by chemically or otherwise introducing a modest number of electrical carriers into an insulating parent compound, thus reaching an unfamiliar and often puzzling regime. The need to understand the properties of this intermediate metal–insulator transition (MIT) region (Mott, 1990) opened an important new Pandora's box, both from the technological and the basic science perspective.

The fundamental question thus arises: What is the physical nature of the MIT phase transition (Dobrosavljević et al., 2012), and what degrees of freedom play a dominant role in its vicinity? The answer is relatively simple in instances where the MIT is driven by an underlying thermodynamic phase transition to an ordered state. This happens, for example, with onset of antiferromagnetic (AFM) order or charge order (CO), which simply modifies the periodic potential experienced by the mobile electrons, and the BTS picture suffices (Slater, 1951). Here, the MIT can simply be regarded as an experimental manifestation of the emerging order, reflecting the static symmetry change of the given material.

The situation is much more interesting in instances where the MIT is not accompanied with some incipient order. Here, the physical reason for a sharp MIT must reside beyond the BTS description, involving either strong electron–electron interactions (Mott, 1949) or disorder (Anderson, 1958). The interplay of both effects proves difficult to theoretically analyze, although considerable effort has been invested, especially within Fermi liquid (FL) (weakly interacting) approaches at weak disorder (Lee and Ramakrishnan, 1985; Finkel'stein, 1983, 1984). More recent work found evidence that disorder may even more significantly modify the properties of strongly correlated metals, sometimes leading to "non-Fermi liquid" metallic behavior (Miranda and Dobrosavljevic, 2005). In any case, the MIT problem in presence of both strong electron–electron interactions and disorder is still far from being fully understood despite recent advances (Dobrosavljević et al., 2012).

On physical grounds, however, one may expect that in specific materials either the correlations or the disorder may dominate, so a simpler theory may suffice. In this chapter we shall not discuss many of the interesting experimental systems where the MIT problem has been investigated so far; a comprehensive overview is given in the chapter by D. Popović. Instead, we shall focus on three specific classes of low-disorder materials showing remarkable similarities in their phenomenology. Our story begins with a short overview of the key experimental features observed in ultraclean diluted two-dimensional electron gases (2DEGs) in semiconductors, displaying the so-called two-dimensional metal–insulator transition (2D-MIT) in a zero magnetic field. Next, we review experiments where very similar behavior is observed for ^{3}He monolayers and also in quasi-2D organic charge transfer salts of the κ-family, both of which are believed to represent model systems displaying the Mott (interaction-driven) MIT. Finally, we provide a direct comparison between these experimental findings and recent theoretical results for interaction-driven (Mott and Wigner–Mott) MITs in absence of disorder. Our message is that the Mott picture represents the dominant physical mechanism in all these systems and should be regarded as the proper starting point for the study of related materials with stronger disorder.

1.2 Phenomenology of 2D-MIT in the Ultraclean Limit

2DEG devices in semiconductors have been studied for almost 50 years since the pivotal 2DEG discovery in 1966 (Fowler et al., 1966). Because they served as basis for most modern electronics, 2DEG systems have been very carefully studied and characterized (Ando et al., 1982) from the materials point of view, allowing the fabrication of ultrahigh-mobility materials by the 1990s. However, their use as an ideal model system for the study of MIT has been recognized relatively late, with the pioneering work of Sergey Kravchenko and collaborators (Kravchenko et al., 1995). The reason was the long-held belief that all electronic states remain localized at $T = 0$, even in the presence of infinitesimally weak disorder, based on the influential scaling theory of localization (Abrahams et al., 1979). This result—the absence of a sharp MIT in two dimensions— was firmly established only for models of noninteracting electrons with disorder, and although even early results on interacting models (Finkel'stein, 1983, 1984) suggested otherwise, Kravchenko's initial work was initially met with extraordinary skepticism and disbelief (Altshuler et al., 2001).

Soon enough, however, clear evidence of a sharp 2D-MIT in high-mobility silicon was confirmed on IBM samples (the original material Kravchenko used came from Russia), with almost identical results (Popović et al., 1997). This marked the beginning of a new era; the late 1990s witnessed a veritable avalanche of experimental and theoretical work on the subject, giving rise to much controversy and debate. This article will not discuss all the interesting results obtained in different regimes, many of which are reviewed in the chapter by D. Popović in this volume. Our focus here will be on documenting evidence that strong electronic correlations—and not disorder—represent the main driving force for 2D-MIT in high-mobility samples. In the following sections, we follow the historical development of the field, with emphasis on those experimental signatures that emerge as robust and rather universal features of interaction-driven MITs in a variety of systems.

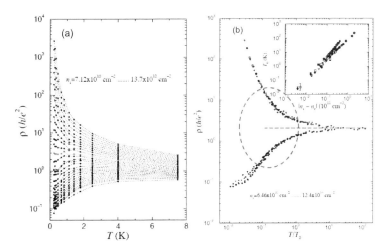

Figure 1.1 (a) Resistivity curves (left panel) for a high-mobility 2DEG in silicon (Kravchenko et al., 1995). (b) Scaling behavior is found over a comparable temperature range. The remarkable mirror symmetry (Simonian et al., 1997a; Dobrosavljević et al., 1997) of the scaling curves (indicated by the dashed oval) seems to hold over more than an order of magnitude for the resistivity ratio. Reprinted (figure) with permission from [Kravchenko, S. V., Mason, W. E., Bowker, G. E., Furneaux, J. E., Pudalov, V. M., and D'Iorio, M. (1995). *Phys. Rev. B* **51**, p. 7038.] Copyright (1995) by the American Physical Society.

1.2.1 Finite-Temperature Transport

2DEG devices provide a very convenient setup for the study of MITs because one can readily control the carrier density in a single device. Thus one easily obtains an entire family of resistivity curves, covering a broad interval of densities and temperatures. Typical results for a high-mobility 2DEG in silicon are shown in Fig. 1.1a, showing a dramatic metal–insulator crossover as the density is reduced below $n_c \approx 10^{11}$ cm^{-2}.

One can observe a rather weak temperature dependence of the resistivity around the critical density (shown as a dashed oval in Fig. 1.1a), with much stronger metallic (insulating) behavior at higher (lower) densities. Note that the system has "made up its mind" whether to be a metal or an insulator even at, relatively speaking, surprisingly high temperatures $T \approx T_F \approx 10$ K. This behavior should be contrasted to that typical of good metals such as silver or gold, where at accessible temperatures $T \ll T_F \approx 10^4$ K. Physically, this means that for 2D-MIT the physical processes relevant for localization already kick in at relatively high temperatures. Here, the system can no longer be regarded as a degenerate electron liquid with dilute elementary excitations. In contrast, simple estimates show (Abrahams et al., 2001) that around the critical density, the characteristic Coulomb energy (temperature) $T_{Coul} \approx 10\ T_F \approx$ 100 K. Clearly, the transition occurs in the regime where Coulomb interactions dominate over all other energy scales in the problem. We should also mention that in the Kelvin range, there essentially are no phonons in silicon (Ando et al., 1982).

We conclude that in the relevant temperature regime, transport should be dominated by inelastic electron–electron scattering, not the impurity-induced (Anderson) localization of quasiparticles, as described by disordered FL theories of Finkel'stein and followers (Finkel'stein, 1983, 1984; Punnoose and Finkel'stein, 2001). As illustrated schematically in Fig. 1.2, transport in this incoherent regime has a very different character than in the coherent "diffusive" regime found in good metals with disorder. Incoherent transport dominated by inelastic electron–electron scattering is typical of systems featuring strong electronic correlations, such as heavy-fermion compounds (Stewart, 1984) or transition metal oxides (TMOs) (Goodenough, 1963) close to the Mott (interaction-driven) MIT (Mott, 1990). Understanding this physical regime requires not only a different set theoretical tools but also an entirely different conceptual picture of electron dynamics. As we shall explain next, modern dynamical mean-field theory (DMFT) methods (Georges et al., 1996) make it possible to qualitatively and even quantitatively explain most universal features within this incoherent transport regime dominated by strong correlation effects.

1.2.2 Scaling Phenomenology and Its Interpretation

There is no doubt that dramatic changes of transport arise in a narrow density range around $n \approx n_c$, but this alone is not enough to mark the existence of a second-order (continuous) phase transition. In particular, all known classical (Goldenfeld, 1992)

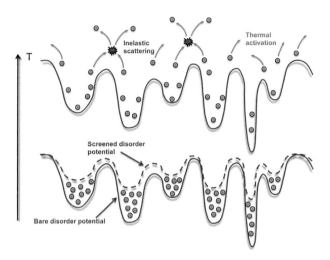

Figure 1.2 In the disordered Fermi liquid picture, the leading low-temperature dependence of transport reflects elastic scattering off a renormalized but temperature-dependent random potential (dashed line). At low temperatures (bottom), the potential wells "fill up" with electrons; in the presence of repulsive (Coulomb) interactions, the screened (renormalized) potential has reduced amplitude (dashed line), leading to effectively weaker disorder. As the temperature increases (top), electrons thermally activate (shown by arrows) out of the potential wells, reducing the screening effect. This physical mechanism, which operates both in the ballistic and in the diffusive regime (Zala et al., 2001), is at the origin of all "quantum corrections" found within the FL picture (Lee and Ramakrishnan, 1985). It is dominant, provided that inelastic electron–electron scattering can be ignored. While this approximation is well justified in good metals, inelastic scattering (star symbol) is considerably enhanced in the presence of strong correlation effects, often leading to disorder-driven non–Fermi liquid behavior (Miranda and Dobrosavljevic, 2005) and electronic Griffiths phases (Andrade et al., 2009).

and even quantum (Sachdev, 2011) critical points also display the characteristic scaling phenomenology. In the case of MITs, one expects (Dobrosavljević et al., 2012) that the resistivity assumes the following scaling form as a function of the reduced density $\delta_n = (n - n_c)/n_c$ and temperature

$$\rho^*(\delta_n, T) = F(T/T_o(\delta_n)), \tag{1.1}$$

where $\rho^*(T) = \rho(\delta_n, T)/\rho(\delta_n = 0, T)$ is the reduced resistivity and $T_o(\delta_n) \approx \delta_n^{\nu z}$ is the corresponding crossover temperature. While similar families of resistivity curves were reported earlier, a big breakthrough for 2D-MIT occurred when Kravchenko demonstrated (Kravchenko et al., 1995) precisely such scaling behavior, providing strong evidence of quantum-critical (QC) behavior expected at a sharply defined $T = 0$ MIT point. It should be stressed, though, that such scaling behavior must arise at any quantum phase transition (QPT), but this alone does not provide direct insight into its physical content or mechanism.

On the other hand, soon after the discovery of 2D-MIT, another interesting feature was noticed (Simonian et al., 1997a), which is not necessarily expected to hold for any general QPT. By carefully examining the form of the relevant scaling function (see Fig. 1.1b), one observes a remarkable mirror symmetry of the two branches such that

$$\rho^*(\delta_n, T) \approx \sigma^*(-\delta_n, T) = [\rho^*(-\delta_n, T)]^{-1}, \tag{1.2}$$

which was found to hold within the entire QC region, for example, at $T > T_o(\delta)$. In practical terms, this means that in the corresponding QC regime the resistivity assumes a stretched-exponential form

$$\rho^*(\delta_n, T) \approx \exp\{A\delta_n/T^x\}, \tag{1.3}$$

where A is a constant and $x = 1/\nu z$. Such exponentially strong temperature dependence is not surprising on the insulating side, where it obtains from activated or hopping transport (Shklovskii and Efros, 1984). In contrast, such inverse activation behavior is highly unusual on the metallic side, where it reflects the dramatic drop of resistivity at low temperatures, at densities just above the transition.

A plausible physical interpretation was quickly proposed (Dobrosavljević et al., 1997) soon after the experimental discovery, based on general scaling considerations. It emphasized that such stretched-exponential form for the relevant scaling function should be interpreted as a signature of strong coupling behavior. Physically, it indicates that the critical region is more akin to the insulating than to the metallic phase. This observation, while having purely phenomenological character, emphasized the key question that one should focus on: the clue to the nature of 2D-MIT should be contained in understanding the physical nature of the insulating state. Is the insulating behavior essentially the result of impurities

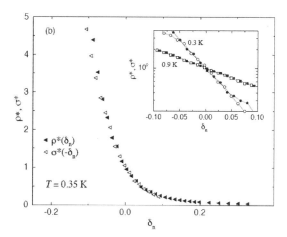

Figure 1.3 Experimental data (Simonian et al., 1997a) displaying mirror symmetry. The inset shows the same data on a semilogarithmic scale, emphasizing linear density dependence of ln ρ consistent with Eq. 1.3, that is, the stretched-exponential phenomenology (Dobrosavljević et al., 1997). Reprinted (figure) with permission from [Simonian, D., Kravchenko, S. V., and Sarachik, M. P. (1997a). *Phys. Rev. B* **55**, p. R13421.] Copyright (1997a) by the American Physical Society.

trapping the mobile electrons, or is the electron–electron repulsion preventing them to move around? If the correlation effects—and not disorder—represent the dominant physical mechanism for localization in other systems, then all the scaling features observed in the 2DEG should again be found. Remarkably, this is exactly what is observed (see later) in very recent experiments on organic Mott systems (Furukawa et al., 2015).

1.2.3 Resistivity Maxima in the Metallic Phase

The QC resistivity scaling has been observed close to the critical density, but other interesting features are also found further out on the metallic side. Here, especially in the cleanest samples, one observes increasingly pronounced resistivity maxima at temperatures $T = T_{max}(\delta)$, which decrease as the transition is approached. Remarkably, the family of curves displaying such maxima persists even relatively deep inside the metallic phase, far outside the critical

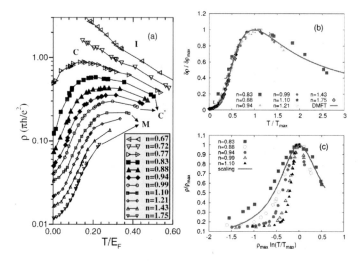

Figure 1.4 (a) Resistivity as a function of temperature from the experiments on a 2DEG in silicon (Pudalov et al., 1998). Only data points indicated by full black symbols have been estimated to lie in the diffusive (coherent) regime, while all metallic curves have an almost identical shape. (b) Same data scaled according to the procedure based on the Wigner–Mott (disorder-free) picture (Radonjić et al., 2010). (c) The scaling procedure based on the disorder-screening scenario (Punnoose and Finkel'stein, 2001) does not produce a convincing collapse. In both scaling plots, the full line is the result of the respective theoretical calculations. Reprinted (figures) with permission from [Radonjić, M. M., Tanasković, D., Dobrosavljević, V., Haule, K., and Kotliar, G. (2012). *Phys. Rev. B* **85**, p. 085133.] Copyright (2012) by the American Physical Society.

region. However, even a quick glance at the experimental data (Fig. 1.4) reveals that all such curves have essentially the same qualitative shape and only differ by the density-dependent values of T_{max} and $\rho_{max} = \rho(T_{max})$.

While this phenomenology is clearly seen, its origin is all but obvious, since resistivity maxima can arise in many unrelated situations. One possibility is the competition of two different disorder-screening mechanisms, as envisioned within the Finkel'stein picture of a disordered FL (Punnoose and Finkel'stein, 2001). However, this mechanism applies (as the authors themselves pointed out) only within the disorder-dominated diffusive regime, which is confined to low temperatures and sufficiently strong impurity scattering. According to reasonable estimates (Punnoose and Finkel'stein,

2001) for the experiment of Fig. 1.4, this picture should only apply to the three curves (full black symbols) close to the critical density, and completely different physics should kick in deeper in the metallic regime (open symbols). This seems unreasonable, since the experimental curves show very similar form at all metallic densities. The key test of these ideas, therefore, should be to examine different materials with strong correlations and weaker disorder, where most of the relevant temperature range is well outside the diffusive regime. If the resistivity maxima, with similar features, persist in such systems, their origin must be of a completely different nature than the disorder-screening mechanism described by the disordered FL picture.

A related issue is the origin of the pronounced resistivity drop at $T < T_{max}$, which can be as large as 1 order of magnitude. Here again very different theoretical interpretations have been proposed, which may or may not apply in a broad class of systems. One issue is the role of impurities (Punnoose and Finkel'stein, 2001), which may be addressed by experimentally examining materials with weak or negligible amounts of disorder. Another curious viewpoint is that of Kivelson and Spivak (Jamei et al., 2005), who proposed that the microscopic phase coexistence (bubble and/or stripe phases) between the Wigner crystal and an FL may explain this behavior, even in the absence of disorder. This possibility may be relevant in some regimes, but certainly not in lattice systems such as heavy-fermion compounds, or systems displaying the (interaction-induced) Mott transition in the absence of any significant disorder.

1.2.4 Effect of Parallel Magnetic Fields

One of the most important clues about 2D-MIT has been obtained soon following its discovery, from studies of magnetotransport. While much of the traditional work on 2DEG systems focuses on the role of perpendicular magnetic fields and the associated quantum Hall regime, for 2D-MIT the application of magnetic fields parallel to the 2D layer sheds important new light. It was found (Simonian et al., 1997b) that the application of parallel magnetic fields can dramatically suppress (Fig. 1.5) the resistivity maxima and the pronounced low-temperature resistivity drop on the metallic side of the transition. This finding is significant because parallel fields

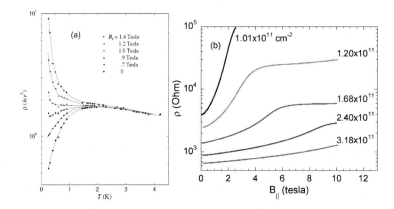

Figure 1.5 (a) Resistivity versus temperature for five different fixed magnetic fields applied parallel to the plane of a low-disordered silicon MOSFET. Here $n = 8.83 \times 10^{10}$ cm^{-2}. Reprinted (figure) with permission from [Simonian, D., Kravchenko, S. V., Sarachik, M. P., and Pudalov, V. M. (1997b). *Phys. Rev. Lett.* **79**, p. 2304.] Copyright (1997b) by the American Physical Society. (b) Low-temperature magnetoresistance as a function of parallel magnetic field at different electron densities above n_c. Reprinted (figure) with permission from [Shashkin, A. A., Kravchenko, S. V., Dolgopolov, V. T., and Klapwijk, T. M. (2001). *Phys. Rev. Lett.* **87**, p. 086801.] Copyright (2001) by the American Physical Society.

couple only to the electron spin (Zeeman splitting), not to the orbital motion of the electron. It points to the important role of spin degrees of freedom in stabilizing the metallic phase.

An interesting fundamental issue relates to the precise role of Zeeman splitting on the ground state of a 2DEG. Is an infinitesimally small value of the parallel field sufficient to suppress the true metal at $T = 0$, or is there a field-driven MIT at some finite value of the parallel field? This important question was initially a subject of much controversy, but careful experimental investigation at ultralow temperatures has established (Shashkin et al., 2001; Eng et al., 2002) the existence of a field-driven MIT (Fig. 1.6) at a finite parallel field but only within a density range close to the $B = 0$ MIT. Interestingly, this field-driven transition appears to belong to a different universality class (Eng et al., 2002) than the zero-field transition (see also the chapter by D. Popović), somewhat similar to what happens in bulk doped semiconductors (Sarachik, 1995).

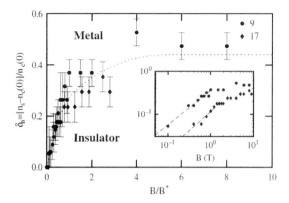

Figure 1.6   Phase diagram for a high-mobility 2DEG in silicon (Eng et al., 2002) as a function of electron density and the parallel magnetic field. The field-driven MIT is found only in the narrow density range $0 < \delta_n < 0.4$, close to the $B = 0$ critical density. Reprinted (figure) with permission from [Eng, K., Feng, X. G., Popović, D., and Washburn, S. (2002). *Phys. Rev. Lett.* **88**, p. 136402.] Copyright (2002) by the American Physical Society.

1.2.5 Thermodynamic Response

The early evidence for the existence of 2D-MIT relied on scaling features of transport in a zero magnetic field (Abrahams et al., 2001), but despite the obvious beauty and elegance of the data, some people (Altshuler et al., 2001) remained skeptical and unconvinced. Complementary insight was therefore of crucial importance, and in the last 15 years this sparked a flurry of experimental work focusing on a thermodynamic response. A detailed account of this large and impressive body of work has been given elsewhere (Kravchenko and Sarachik, 2004); it is also covered in the chapter by Shashkin and Kravchenko. Several outstanding features have been established by these experiments, as follows.

(1) The spin susceptibility is strongly enhanced at low temperatures, and it seems to evolve (Prus et al., 2003) from a Pauli-like form at higher densities (typical for metals) to a Curie-like form (expected for localized magnetic moments) as the transition is approached. This result indicates that most electrons convert into spin-1/2 localized magnetic moments within the insulating state.

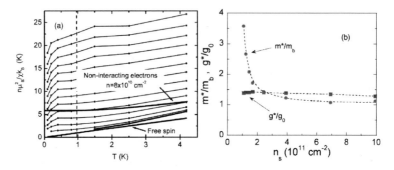

Figure 1.7 (a) Inverse spin susceptibility normalized per carrier, as obtained from magnetocapacitance experiments (Prus et al., 2003). Experimental curves from bottom to top correspond to densities $0.8 - 6 \times 10^{11}$ cm^{-2} in 4×10^{10} cm^{-2} steps. The thick straight line depicts the Curie law (slope given by the Bohr magneton), and the dashed line marks $T = (g_B/k_B)0.7T$. Reprinted (figure) with permission from [Prus, O., Yaish, Y., Reznikov, M., Sivan, U., and Pudalov, V. (2003). *Phys. Rev. B* **67**, p. 205407.] Copyright (2003) by the American Physical Society. (b) Renormalization of the effective mass (dots) and the g-factor (squares), as a function of electron density. Reprinted (figure) with permission from [Shashkin, A. A., Kravchenko, S. V., Dolgopolov, V. T., and Klapwijk, T. M. (2002). *Phys. Rev. B* **66**, p. 073303.] Copyright (2002) by the American Physical Society.

(2) Various complementary experimental methods have firmly established (Kravchenko and Sarachik, 2004) pronounced interaction-induced enhancement of the (quasiparticle) effective mass m^*, which appears to linearly diverge at $n = n_c$. In contrast, the corresponding g-factor does not display any significant renormalization. This important observation rules out incipient ferromagnetism as a possible origin of enhanced spin susceptibility. Instead, it points to local magnetic moment formation within the insulating phase. These earlier findings were recently confirmed by spectacular thermopower measurements, where values of m^*/m as large as 25 were reported (Mokashi et al., 2012).

(3) The vast majority of the relevant experiments in this category have been performed within the ballistic regime. This is significant because the predictions of the disordered FL picture of Finkel'stein (Punnoose and Finkel'stein, 2001) apply only

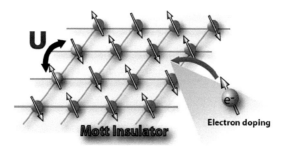

Figure 1.8 In a Mott insulator, each electron resides in a bound electronic state, forming a spin-1/2 local magnetic moment. The lowest energy charge excitation costs an energy comparable to the on-site Coulomb repulsion U.

within the diffusive regime and thus cannot provide any insight into the origin of these thermodynamic anomalies.

(4) All these thermodynamic signatures are best seen in the cleanest samples, and they display surprisingly weak dependence on the sample mobility (level of disorder). This again points to strong correlations as the dominant mechanism in this regime.

1.3 Comparison to Conventional Mott Systems

Experiments on 2DEG systems have suggested that 2D-MIT should best be viewed as an interaction-driven MIT, where the insulator consists of localized magnetic moments. To put these ideas in perspective, it is useful to compare these findings to the properties of other 2D systems known to display the Mott (interaction-driven) MIT (Mott, 1949, 1990).

Mott insulators often display some form of magnetic order at low temperatures, but in contrast to the band picture (Slater, 1934), their insulating nature is not tied down to magnetic order. Indeed, a substantial gap in charge transport is typically seen even at temperatures much higher than the magnetic ordering temperature. Such a QPT (Dobrosavljević et al., 2012) between a paramagnetic FL metal and a paramagnetic Mott insulator—the "pure" Mott transition—is our main focus here. In particular, we shall focus on two specific examples of the broad family of Mott materials, where

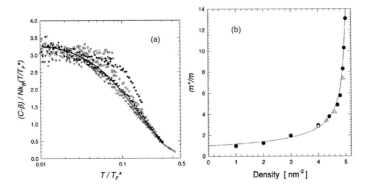

Figure 1.9 Approaching the Mott transition in ^{3}He monolayers on graphite. (a) Reduced fluid heat capacity as a function of reduced temperature, showing emergence of a Fermi liquid at low temperatures. Here $T_F^* = T_F/(m^*/m)$ is the quasiparticle Fermi temperature. (b) Effective mass ratio as a function of ^{3}He fluid density inferred from heat capacity (full circle) and magnetization (open triangles), showing apparent divergence. The fact that both quantities display the same critical behavior indicates that $g^* \approx$ const. Reprinted (figure) with permission from [Casey, A., Patel, H., Ny' eki, J., Cowan, B. P., and Saunders, J. (2003). *Phys. Rev. Lett.* **90**, p. 115301.] Copyright (2003) by the American Physical Society.

it proved possible to experimentally study several features with striking similarity to the behavior found in a 2DEG displaying 2D-MIT.

1.3.1 Mott Transition in ^{3}He Monolayers on Graphite

The ^{3}He liquid consists of charge-neutral spin-1/2 atoms with Fermi statistics; its normal phase at ambient pressure is often regarded as a model system for a strongly correlated FL (Vollhardt, 1984). This includes significant enhancements of the specific heat and the spin susceptibility, reflecting the emergence of heavy quasiparticles (Pines and Nozières, 1965), as first predicted by Landau (Landau, 1957). ^{3}He will solidify under increasing hydrostatic pressure, due to strong interatomic repulsion at short distances. Each spin-1/2 He^3 atom then sits on a different lattice site, thus forming a Mott insulator.

The associated solidification transition has long been viewed as a realization of the Mott MIT. Many of its low-temperature properties, on the fluid side, can be described (Vollhardt, 1984) by the variational solution of the Hubbard model based on the Gutzwiller approximation (Brinkman and Rice, 1970), predicting the divergence of the effective mass as

$$m^*/m \approx (P_c - P)^{-1}. \qquad (1.4)$$

However, similarly as in most other freezing transitions, bulk ^{3}He displays a rather robust first-order solidification phase transition associated with the emergence of ferromagnetic (FM) order within the crystalline phase. Consequently, one is able to observe only a restricted range of m^* values, making it difficult to compare to theory. Despite these limitations, many qualitative and even quantitative features can be understood from this perspective, including the findings that application of strong magnetic fields may trigger the destruction of the FL state (Vollhardt, 1984) and the associated field-induced solidification of ^{3}He.

More recent work by Saunders and collaborators (Casey et al., 2003) has focused on ^{3}He fluid monolayers placed on graphite, where again one can study the approach to the Mott transition by controlling hydrostatic pressure. Most remarkably, the first-order jump at solidification is then suppressed, presumably because the solid ^{3}He monolayer on graphite displays no magnetic order down to $T = 0$. Instead, nonmagnetic spin liquid behavior is found within the solid phase, due to important ring exchange processes in this 2D triangular lattice (Misguich et al., 1998). The corresponding Mott transition is found to be of second order, with rather spectacular agreement with the Brinkmann–Rice (BR) theory for the m^* divergence. According to FL theory (Pines and Nozières, 1965), the Sommerfeld coefficient $\gamma = C/T \approx m^*$ depends only on the effective mass, while the spin susceptibility $\chi \approx m^*/(1 + F_o^a) = m^* g^*$ also involves the (renormalized) g-factor. This is significant because the behavior of the g-factor distinguishes the BR (Mott) transition from an FM instability, although χ diverges in both instances. Namely, g^* is expected to diverge at the FM transition (Pines and Nozières, 1965), while it should remain constant at the BR point (Vollhardt, 1984).

To address this important issue, it suffices to determine both m^* and g^* or, equivalently, to measure both γ and χ. Precisely such an analysis was carried out for ^{3}He monolayers (Casey et al., 2003), where an m^* enhancement as large as 15 was reported, while g^* was found to remain constant in the critical region and assume a value close to the BR prediction. This finding is strikingly similar to what is observed in analogous studies (Kravchenko and Sarachik, 2004) for ultraclean 2DEG systems close to 2D-MIT; it suggests that both phenomena may have a common physical origin: the behavior of a strongly correlated FL in the proximity of the Mott insulating phase.

Transport properties are more difficult to probe in ^{3}He, because its charge neutrality precludes direct coupling of electric fields to density currents. One cannot, therefore, easily compare its transport properties to that of 2DEG systems in order to further test the Mott picture. To do this, we next turn to other classes of canonical Mott systems where recent work has provided important insights. Over the last 30 years, many studies of the Mott transition have focused on various TMOs (Goodenough, 1963), especially after the discovery of high-temperature superconductivity in the late 1980s. These materials, however, are notorious for displaying all kinds of problems in material preparation and sample growth and often contain substantial amounts of impurities and defects. Another class of systems where better sample quality is somewhat easier to obtain is the so-called organic Mott systems, which we discuss next.

1.3.2 Mott Organics

The organic Mott systems have emerged, over 30 years, as a very popular set of materials for the study of the Mott MIT. These are organic charge transfer salts consisting of a quasi-2D lattice of rather large organic molecules, with typically a single molecular orbital close to the Fermi energy. Since the intermolecular overlap of such orbitals is typically modest, these crystals usually have very narrow electronic bands with substantial intraorbital (on-site) Coulomb repulsion and hence are well described by 2D single-band Hubbard-type models (Powell and McKenzie, 2011).

There exist two general families of organic Mott crystals, the κ-family, corresponding to a half-filled Hubbard model, and the

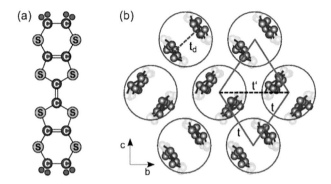

Figure 1.10 Sketch of the BEDT-TTF molecule (Elsässer et al., 2012). For κ-(BEDT-TTF)$_2$X, the molecules are arranged in dimers, which constitute an anisotropic triangular lattice within the conduction layer. Reprinted (figure) with permission from [Elsässer, S., Wu, D., Dressel, M., and Schlueter, J. A. (2012). *Phys. Rev. B* **86**, p. 155150.] Copyright (2012) by the American Physical Society.

θ-family, corresponding to quarter filling. The former class is particularly useful for the study of the bandwidth-driven Mott transition at half filling, since the electronic bandwidth can be conveniently tuned via hydrostatic pressure. A notable feature of these materials is a substantial amount of magnetic frustration, due to nearly isotropic triangular lattices formed by the organic molecules within each 2D layer. As a result, several members of this family (such as κ-(BEDT-TTF)$_2$Cu$_2$(CN)$_3$) display no magnetic order down to the lowest accessible temperatures, while others (such as κ-(BEDT-TTF)$_2$Cu[N(CN)$_2$]Cl) feature an AFM order in the insulating phase. Another important property of these materials is the excellent quality of crystals, with very little disorder, making the dominance of strong correlation effects obvious in all phenomena observed.

On the metallic side, in all these materials display well charac-terized FL behavior, with substantial effective mass enhancements, similarly to other Mott systems. In contrast to the conventional QC scenario, however, here the phase diagram features a first-order MIT and the associated coexistence region, terminating at the critical end point $T = T_c$. Still, as emphasized in recent theoretical (Terletska et al., 2011) and experimental (Furukawa et al., 2015) work, the

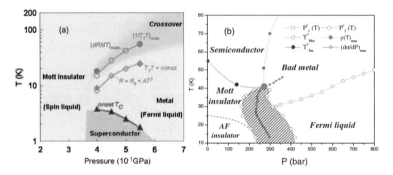

Figure 1.11 Phase diagrams of (a) κ-(BEDT-TTF)$_2$Cu$_2$(CN)$_3$ displaying no magnetic order. Reprinted (figure) with permission from [Kurosaki, Y., Shimizu, Y., Miyagawa, K., Kanoda, K., and Saito, G. (2005). *Phys. Rev. Lett.* **95**, p. 177001.] Copyright (2005) by the American Physical Society. and (b) κ-(BEDT-TTF)$_2$Cu[N(CN)$_2$]Cl featuring an AFM order (Limelette et al., 2003) on the insulating side. Reprinted (figure) with permission from [Limelette, P., Wzietek, P., Florens, S., Georges, A., Costi, T. A., Pasquier, C., Jerome, D., Meziere, C., and Batail, P. (2003). *Phys. Rev. Lett.* **91**, p. 016401.] Copyright (2003) by the American Physical Society.

temperature range associated with this Mott coexistence region is typically very small ($T_c \approx 20$–40 K) in comparison to the Coulomb interaction or the (bare) Fermi temperature ($T_F \approx 2000$ K). Note that the comparable temperature range in 2DEG systems would be extremely small, viz., $T < 10^{-2}\, T_F \approx 0.1$ K, while most experimental results (especially those obtained in the ballistic regime) correspond to much higher temperatures.

We should emphasize that different materials in this class display various magnetic or even superconducting orders within the low-temperature regime $T < T_c$ (Fig. 1.11), which strongly depend on material details, such as the precise amount of magnetic frustration. In contrast, the behavior above the coexistence dome is found to be remarkably universal, featuring very similar behavior in all compounds. Here, one finds a smooth crossover between an FL metal and a Mott insulator, displaying a family of resistivity curves with a general form very similar to those found in 2DEG systems close to 2D-MIT, in a temperature range comparable to a fraction of T_F. In addition to a fan-like shape at high temperatures, we note the existence of resistivity maxima on the metallic side, which become

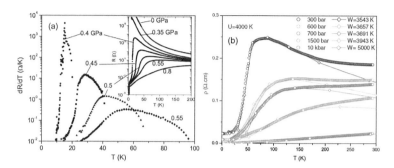

Figure 1.12 Transport in Mott organic materials. (a) κ-(BEDT-TTF)$_2$ Cu$_2$(CN)$_3$. Reprinted (figure) with permission from [Kurosaki, Y., Shimizu, Y., Miyagawa, K., Kanoda, K., and Saito, G. (2005). *Phys. Rev. Lett.* **95**, p. 177001.] Copyright (2005) by the American Physical Society. (b) κ-(BEDT-TTF)$_2$Cu[N(CN)$_2$]Cl. The lines are obtained from DMFT. Reprinted (figure) with permission from [Limelette, P., Wzietek, P., Florens, S., Georges, A., Costi, T. A., Pasquier, C., Jerome, D., Meziere, C., and Batail, P. (2003). *Phys. Rev. Lett.* **91**, p. 016401.] Copyright (2003) by the American Physical Society.

more pronounced closer to the transition, precisely as in many 2DEG examples.

In addition, very recent experimental work (Furukawa et al., 2015) performed a careful scaling analysis of the resistivity data in the high-temperature region ($T > T_c$), providing clear and convincing evidence of QC behavior associated with the Mott point. The analysis was performed on three different materials (including two compounds discussed above), featuring very different magnet ground states (AFM order vs. spin liquid behavior) on the insulating side. Nevertheless, the scaling analysis succeeded in extracting the universal aspects of transport in this QC regime, finding impressive universality (Fig. 1.13) both in the form of the corresponding scaling function and in the values of the critical exponents, with excellent agreement with earlier DMFT predictions (Terletska et al., 2011).

Most remarkably, the experimentally determined scaling functions display precisely the same mirror symmetry as earlier found on 2DEG systems (Simonian et al., 1997a), with exactly the stretched-exponential form proposed by phenomenological scaling theory (Dobrosavljević et al., 1997). Given the fact that 2DEG systems and these organic Mott crystals have completely different

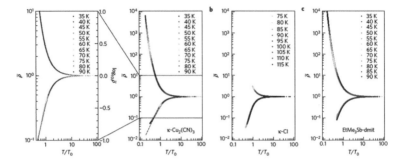

Figure 1.13 Quantum-critical scaling of the resistivity curves for three different organic Mott materials, demonstrating (Furukawa et al., 2015) universal behavior insensitive to a low-temperature magnetic order. Reprinted by permission from Macmillan Publishers Ltd: [*Nature Physics*] (Furukawa, T., Miyagawa, K., Taniguchi, H., Kato, R., and Kanoda, K. (2015). *Nat. Phys.* **11**, pp. 221–224), copyright (2015).

microscopic character, these experimental findings provide direct and clear evidence of a surprisingly robust universality of the QC behavior in interaction-driven MITs. If one and the same basic physical processes—Mott localization—indeed dominate all these phenomena, then one can hope that even a theory based on simple models may account for the main trends.

1.4 Theory of Interaction-Driven MITs

Interaction-driven MITs in the absence of static symmetry breaking—the Mott transition—have first proposed (Mott, 1949) as a possible mechanism in doped semiconductors. Soon enough, it became apparent that they are the right physical picture for many systems at the brink of magnetism, for example, various TMOs (Mott, 1990; Goodenough, 1963). Such systems are typically characterized by a narrow valence band close to half filling, with a substantial on-site (intra-orbital) Coulomb interaction U. The simplest generic model describing this situation is the single-band Hubbard model given by the Hamiltonian

$$H = -t \sum_{\langle ij \rangle \sigma} c_{i\sigma}^{\dagger} c_{j\sigma} + U \sum_{i} n_{i\uparrow} n_{i\downarrow}. \qquad (1.5)$$

Here, t is the hopping amplitude, $c_{i\sigma}^{\dagger}$ ($c_{j\sigma}$) are the creation (annihilation) operators, $n_i = n_{i\uparrow} + n_{i\downarrow}$ is the occupation number operator on site i, and U is the on-site Coulomb interaction.

Theoretical investigations of the Hubbard model go back to the pioneering investigations in the early 1960s (Hubbard, 1963) when it already became clear that, even in the absence of any magnetic order, a Hubbard–Mott gap will open for a sufficiently large U/t. Approaching the transition from the metallic side was predicted (Brinkman and Rice, 1970) to result in substantial effective mass enhancement as a precursor of the Mott transition, as confirmed in numerous TMOs and other materials. A reliable description of the transition region, however, remained elusive for many years. Still, the existence a broad class of Mott insulators, containing localized magnetic moments, has long been regarded as being well established (Goodenough, 1963).

1.4.1 The DMFT Approach

Modern theories of Mott systems were sparked by the discovery of high-temperature superconductors in the late 1980s, which triggered considerable renewed interest in the physics of TMOs and related materials. A veritable zoo of physical ideas and theoretical approaches was put forward, but few survived the test of time. Among those, DMFT (Georges et al., 1996) proved to be the most useful, reliable, and flexible tool, applicable both in modeling

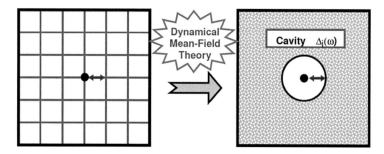

Figure 1.14 In dynamical mean-field theory, the environment of a given site is represented by an effective medium represented by its cavity spectral function $\Delta_i(\omega)$.

Hamiltonians but also in first-principles theories (Kotliar et al., 2006) of correlated matter.

In its simplest (single-site) implementation, DMFT provides a local approximation for many-body corrections in strongly interacting systems. The central quantity it self-consistently calculates is the single-particle self-energy $\Sigma_i(\omega, T)$, which is obtained within a local approximation. Its real part describes the interaction (or temperature) dependence of the electronic spectra, allowing substantial narrowing of the quantum phase (QP) bandwidth (i.e., the effective mass m^* enhancement) and the eventual opening of the Mott gap. Its imaginary part, on the other hand, describes electron–electron (inelastic) scattering processes, leading to the ultimate destruction of the coherent QPs at sufficiently large T or U.

Most importantly, DMFT does not suffer from limitations of standard FL approaches, which are largely restricted to situations with dilute QPs (i.e., the lowest temperatures). In fact, DMFT is most reliable at high temperatures, within the incoherent regime (with large inelastic scattering), where its local approximation becomes essentially exact. Indeed, systematic (nonlocal) corrections to single-site DMFT have established (Tanasković et al., 2011) that there exists, in general, a well-defined crossover temperature scale $T^*(U)$, above which the nonlocal corrections are negligibly small. This scale is typically much smaller than the basic energy scale of the problem (e.g., the bandwidth or the interaction U), so DMFT often proves surprisingly accurate over much of the experimentally relevant temperature range.

Another important aspect of the DMFT approach deserves special emphasis. In contrast to conventional (Slater-like) theories focusing on effects of various incipient orders, DMFT focuses on dynamical correlation effects unrelated to orders. From the technical perspective, its local character suppresses the spatial correlations associated with magnetic, charge, or structural correlations. Physically, such an approximation becomes accurate in presence of sufficiently strong frustration effects due to competing interactions or competing orders, a situation that is typical for most correlated electronic systems. Frustration effects generally lead to a dramatic proliferation of low-lying excited states, which further invalidates the conventional FL picture of dilute elementary excitations; this situation demands an accurate description of incoherent transport regimes—precisely what is provided by DMFT.

Historically, since its discovery more than 20 years ago, DMFT has been applied to many models and various physical systems. Recent developments have also included the application to inhomogeneous electronic systems in the presence of disorder and were able to incorporate the relevant physical processes such as Anderson localization and even the glassy freezing of electrons (the description of quantum Coulomb glasses). These interesting developments have been reviewed elsewhere (Dobrosavljević et al., 2012; Miranda and Dobrosavljevic, 2005) but will not be discussed here. In the following, we briefly review the key physical results of DMFT when applied to the simplest model of a Mott transition at half filling and the associated Wigner–Mott transition away from half filling.

1.4.2 The Mott Transition

Many insulating materials have an odd number of electrons per unit cell; thus band theory would predict them to be metals—in contrast to experiments. Such compounds (e.g., TMOs) often have AFM ground states, leading Slater to propose that spin density wave formation (Slater, 1951) is likely at the origin of the insulating behavior. This mechanism does not require any substantial modification of the band theory picture, since the insulating state is viewed as a consequence of a band gap opening at the Fermi surface.

According to Slater (Slater, 1951), such insulating behavior should disappear above the Neel temperature, which is typically in the 10^2 K range. Most remarkably, in most AFM oxides, clear signatures of insulating behavior persist at temperatures well above any magnetic ordering, essentially ruling out Slater's weak coupling picture.

What goes on in such cases was first clarified in early works by Mott (Mott, 1949) and Hubbard (Hubbard, 1963), tracing the insulating behavior to strong Coulomb repulsion between electrons occupying the same orbital. When the lattice has integer filling per unit cell, then electrons can be mobile only if they have enough

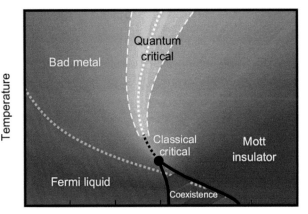

Interaction

Figure 1.15 Phase diagram of the half-filled Hubbard model calculated from DMFT. At very low temperatures $T < T_c \leq 0.03 T_F$ the Fermi liquid and the Mott insulating phases are separated by a first-order transition line and the associated coexistence region. Very recent work (Terletska et al., 2011; Vučičević et al., 2013, 2015) has established that in a very broad intermediate-temperature region $T_c < T < T_F$, one finds characteristic metal–insulator crossover behavior showing all features expected of quantum criticality. Reprinted (figure) with permission from [Vučičević, J., Terletska, H., Tanasković, D., and Dobrosavljević, V. (2013). *Phys. Rev. B* **88**, p. 075143.] Copyright (2013) by the American Physical Society.

kinetic energy ($E_K \approx t$) to overcome the Coulomb energy U. In the narrow band limit of $t \ll U$, the electrons do not have enough kinetic energy, and a gap opens in the single-particle excitation spectrum, leading to Mott insulating behavior. This gap $E_g \approx U - B$ (here $B \approx 2zt$ is the electronic bandwidth; z being the lattice coordination number) is the energy an electron has to pay to overcome the Coulomb repulsion and leave the lattice site.

In the ground state, each lattice site is singly occupied, and the electron occupying it behaves as a spin-1/2 local magnetic moment. These local moments typically interact through magnetic superexchange interactions (Anderson, 1959) of the order $J \approx t^2/U$, leading to magnetic ordering at temperatures of the order T. The insulating behavior, however, is not caused by magnetic ordering and typically persists all the way to temperatures $T \approx E_g \gg T_J$.

In oxides, $E_g \approx 10^3$–10^4 K is typically on the atomic (eV) scale, while magnetic ordering emerges at temperatures roughly an order of magnitude lower $T_J \approx 100$–300 K.

1.4.3 Correlated Metallic State

An important step in elucidating the approach to the Mott transition from the metallic side was provided by the pioneering work of Brinkmann and Rice (Brinkman and Rice, 1970). This work, which was motivated by experiments on the normal phase of ^{3}He (Vollhardt, 1984), predicted a strong effective mass enhancement close to the Mott transition. In the original formulation, as well as in its subsequent elaborations (slave boson mean-field theory (Kotliar and Ruckenstein, 1986), DMFT (Georges et al., 1996), the effective mass is predicted to continuously diverge as the Mott transition is approached from the metallic side

$$\frac{m^*}{m} \approx (U_c - U)^{-1}. \tag{1.6}$$

A corresponding coherence (effective Fermi) temperature

$$T^* \approx T_F/m^* \tag{1.7}$$

is predicted above which the quasiparticles are destroyed by thermal fluctuations. As a result, one predicts (e.g., within DMFT) a large resistivity increase around the coherence temperature and a crossover to insulating (activated) behavior at higher temperatures. Because the low-temperature FL is a spin singlet state, a modest magnetic field of the order

$$B^* \approx T^* \approx (m^*)^{-1} \tag{1.8}$$

is expected to also destabilize such a correlated metal and lead to a large and positive magnetoresistance. Note that both T^* and B^* are predicted to continuously vanish as the Mott transition is approached from the metallic side, due to the corresponding m^* divergence. Precisely such behavior is found in all the three classes of physical systems we discussed above.

1.4.4 Effective Mass Enhancement

How should we physically interpret the large effective mass enhancement which is seen in all these systems? What determines its magnitude if it does not actually diverge at the transition? An answer to this important question can be given using a simple thermodynamic argument which does not rely on any particular microscopic theory or a specific model. In the following section, we present this simple argument for the case of a clean FL, although its physical context is, of course, much more general.

In any clean FL (Pines and Nozières, 1965) the low-temperature specific heat assumes the leading form

$$C(T) = \gamma T + \cdots ,\tag{1.9}$$

where the Sommerfeld coefficient

$$\gamma \approx m^*.\tag{1.10}$$

In the strongly correlated limit $(m^*/m \gg 1)$ this behavior is expected only at $T \lesssim T^* \approx (m^*)^{-1}$, while the specific heat should drop to much smaller values at higher temperatures where the quasiparticles are destroyed. Such behavior is indeed observed in many systems showing appreciable mass enhancements.

On the other hand, from general thermodynamic principles, we can express the entropy as

$$S(T) = \int_0^T dT \, \frac{C(T)}{T}.\tag{1.11}$$

Using the above expressions for the specific heat, we can estimate the entropy around the coherence temperature

$$S(T^*) \approx \gamma T^* \approx O(1).\tag{1.12}$$

The leading effective mass dependence of the Sommerfeld coefficient γ and that of the coherence temperature T^* cancel out!

Let us now explore the consequences of the assumed (or approximate) effective mass divergence at the Mott transition. As $m^* \longrightarrow \infty$, the coherence temperature $T^* \longrightarrow 0+$, resulting in large residual entropy

$$S(T \longrightarrow 0+) \approx O(1).\tag{1.13}$$

We conclude that the effective mass divergence indicates the approach to a phase with finite residual entropy!

Does not this result violate the third law of thermodynamics? And how can it be related to the physical picture of the Mott transition? The answer is, in fact, very simple. Within the Mott insulating phase the Coulomb repulsion confines the electrons to individual lattice sites, turning them into spin-1/2 localized magnetic moments. To the extend that we can ignore the exchange interactions between these spins, the Mott insulator can be viewed as a collection of free spins with large residual entropy $S(0+) = R \ln 2$. This is precisely what happens within the BR picture; similar results are obtained from DMFT, a result that proves exact in the limit of large lattice coordination (Georges et al., 1996).

In reality, the exchange interactions between localized spins always exist, and they generally lift the ground-state degeneracy, restoring the third Law. This happens below a low-temperature scale T_J, which measures the effective dispersion of intersite magnetic correlations (Moeller et al., 1999; Park et al., 2008) emerging from such exchange interactions. In practice, this magnetic correlation temperature T_J can be very low, either due to effects of geometric frustration or due to additional ring exchange processes which lead to competing magnetic interactions. Such a situation is found both in organic Mott systems, and for 2D Wigner crystals, where the insulating state corresponds to a geometrically frustrated triangular lattice. In addition, numerical calculations by Ceperley and others (Cândido et al., 2004) have established that for 2D Wigner crystals significant ring exchange processes indeed provide additional strong frustration effects, further weakening the intersite spin correlation effects.

We conclude that the effective mass enhancement, whenever observed in experiment, indicates the approach to a phase where large amounts of entropy persist down to very low temperatures. Such situations very naturally occur in the vicinity of the Mott transition, since the formation of local magnetic moments on the insulating side gives rise to large amounts of spin entropy being released at very modest temperatures. A similar situation is routinely found (Stewart, 1984; Flouquet, 2005; Hewson, 1993) in the so-called heavy-fermion compounds (e.g., rare-earth intermetallics)

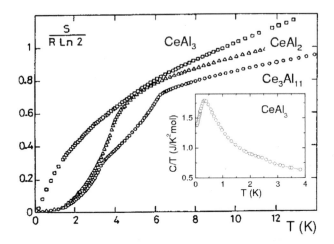

Figure 1.16 Temperature dependence of entropy extracted from specific heat (inset) experiments (Flouquet, 2005) on several heavy-fermion materials. Essentially the entire doublet entropy $S = R \ln 2$ is recovered by the time the temperature has reached $T^* \approx 10$ K, consistent with a large mass enhancement $m^* \approx 1/T^*$. Reprinted from Flouquet, J., Lasjaunias, J. C., Peyrard, J., and Ribault, M. (1982). *J. Appl. Phys.* **53**, pp. 2127–2130. With the permission from AIP Publishing.

featuring huge effective mass enhancements. Here, local magnetic moments coexist with conduction electrons giving rise to the Kondo effect, which sets the scale for the FL coherence temperature $T^* \approx 1/m^*$, above which the entire free spin entropy $S(T^*) \approx R \ln 2$ is recovered (Fig. 1.16). This entropic argument is, in turn, often used to experimentally prove the very existence of localized magnetic moments within a metallic host.

We should mention that other mechanisms of effective mass enhancement have also been considered. General arguments (Millis, 1993) indicate that m^* can diverge when approaching a quantum-critical point (QCP) corresponding to some (magnetically or charge) long-range ordered state. This effect is, however, expected only below an appropriate upper critical dimension (Sachdev, 2011), reflecting an anomalous dimension of the incipient ordered state. In addition, this is a mechanism dominated by long-wavelength order-parameter fluctuations and is thus expected to contribute only a small amount of entropy per degree of freedom in contrast to local moment formation.

It is interesting to mention that weak coupling approaches, such as the popular on-shell interpretation of the random phase approximation (RPA) (Ting et al., 1975), often result in inaccurate or even misleading predictions (Zhang and Das Sarma, 2005) for the effective mass enhancement behavior. Indeed, more accurate modern theories such as DMFT can be used (Dobrosavljević et al., 2012) to benchmark these and other weak coupling theories and to reveal the origin of the pathologies resulting from their inappropriate applications to strong coupling situations.

1.4.5 Resistivity Maxima

Within the strongly correlated regime close to the Mott transition, transport is carried by heavy quasiparticles. As temperature increases, inelastic electron–electron scattering kicks in, leading to a rapid increase of resistivity with temperature. Within FL theory, this gives rise to the temperature dependence of the form $\rho(T) \approx AT^2$, with $A \approx (m^*)^2$ according to the Kadowaki–Woods law (Hewson, 1993). Such behavior is indeed observed in most known correlated system, in agreement with DMFT predictions (Jacko et al., 2009). This argument makes it clear that increasingly strong temperature dependence emerges in the correlated regime, but it does not tell us what happens above the corresponding QP coherence temperature $T^* \approx 1/m^*$.

In this regime inaccessible to conventional FL theories, DMFT provided an illuminating answer (Radonjić et al., 2012). It described the thermal destruction of QPs, and the consequent opening of a Mott pseudogap at $T > T^*$, leading to resistivity maxima. While much of the earlier DMFT work focussed on the low-T regime, more recent advances (Gull et al., 2011) in quantum Monte Carlo methods (needed to solve the DMFT equations) allowed a careful and precise characterization of this transport regime. One finds a family of curves displaying pronounced resistivity maxima with increasing height at temperatures that decrease close to the transition.

These curves all assume essentially the same functional form (Fig. 1.17) and therefore can all be collapsed on a single scaling

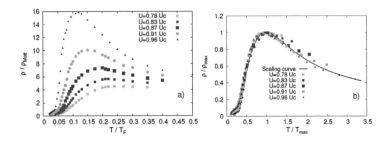

Figure 1.17 (a) Resistivity maxima in the strongly correlated metallic regime of a half-filled Hubbard model within DMFT (Radonjić et al., 2012). (b) Scaling collapse of the resistivity maxima. Reprinted (figure) with permission from [Radonjić, M. M., Tanasković, D., Dobrosavljević, V., Haule, K., and Kotliar, G. (2012). *Phys. Rev. B* **85**, p. 085133.] Copyright (2012) by the American Physical Society.

function by rescaling the temperature with that of the resistivity maximum $T_{\max}$ and the resistivity with its maximum value $\rho_{\max}$. From these numerical results one can extract a universal scaling function describing the entire family of curves (as shown by the thick red line in Fig. 1.17). A direct comparison can now be made with data obtained from experiments on a 2DEG in silicon close to 2D-MIT; one finds surprisingly good agreement between theory and experiment, with no adjustable parameters (see Fig. 1.4). Similar data are found also from experiments on Mott organics and other conventional Mott systems, where successful comparison with DMFT has already been established (Radonjić et al., 2010; Limelette et al., 2003).

Further comparison between experiment and theory is obtained by plotting $T_{\max}$ vs. m^*, where both quantities can be independently obtained both from DMFT and from 2DEG experiments. Here we express $T_{\max}$ in units of the Fermi temperature and m^* in the units of the known band mass in order to perform a comparison with no adjustable parameters. Again, one finds excellent agreement between DMFT and experiments, giving

$$T_{\max}/T_{\mathrm{F}} \approx 0.7(m/m^*), \qquad (1.14)$$

where even the numerical prefactor close to 0.7 is obtained in both cases.

A clear physical understanding of the Mott transition within DMFT, together with spectacular agreement between DMFT and

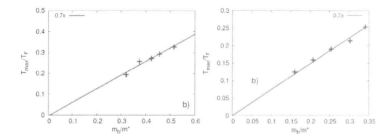

Figure 1.18 Dependence of T_{max} on m^* from (a) DMFT and (b) experiments on a 2DEG close to 2D-MIT. Reprinted (figure) with permission from [Radonjić, M. M., Tanasković, D., Dobrosavljević, V., Haule, K., and Kotliar, G. (2012). *Phys. Rev. B* **85**, p. 085133.] Copyright (2012) by the American Physical Society.

2DEG experiments, paints a convincing and transparent picture of the mechanism for the resistivity maxima, as follows:

- Heavy quasiparticles that exist near the Mott transition are characterized by a small coherence temperature $T^* \approx 1/m^*$.
- Given that both theory and experiments find $T_{max} \approx 1/m^*$, we can directly identify the resistivity maxima with the thermal destruction of heavy quasiparticles.
- The above physical picture is not only well established within DMFT but is also well documented in all strongly correlated electronic systems such as various TMO materials, organic Mott systems, and even heavy-fermion compounds.
- The complete thermal destruction of quasiparticles around $T \approx T_{max}$ is a physical picture completely different than the situation described by the disordered FL scenario (Punnoose and Finkel'stein, 2001), which does not even provide a good data collapse. In this scenario, the relevant quasiparticle regime extends both below and above T_{max}, and the maxima result from the competition of two different elastic (but temperature-dependent) scattering mechanisms (Zala et al., 2001).
- We have seen that, both in DMFT and in all known Mott systems, the metal–insulator coexistence region does exist close to the MIT, but it remains confined to very low

temperatures, typically 2 orders of magnitude smaller than the Fermi energy or the Coulomb repulsion U.

- The resistivity maxima, in contrast, are found at much higher temperatures, typically as high as a 10%–20% of the Fermi temperature. This feature is clearly found both in DMFT, in all conventional Mott systems, but also in 2DEG materials close to 2D-MIT.

- In contrast to DMFT and the known behavior of many conventional Mott systems, the phase separation scenario proposed by Kivelson and Spivak (Jamei et al., 2005) can hold only within the metal–insulator coexistence region. It therefore appears very unconvincing as a possible mechanism behind the resistivity maxima found in 2DEG systems.

1.4.6 Quantum Criticality and Scaling

What is the physical nature of the Mott transition? The physical picture of Brinkmann and Rice (Brinkman and Rice, 1970) makes it plausible that the Mott transition should be viewed as a QCP, since the characteristic energy scale of the correlated FL $T^* \approx T_F/m^*$ continuously decreases as the transition is approached. On the other hand, the Mott transition discussed here describes the opening of a correlation-induced spectral gap in the absence of magnetic ordering, that is, within the paramagnetic phase. Such a phase transition is not associated with spontaneous symmetry breaking associated with any static order parameter. Why should the phase transition then have any second-order (continuous) character at all?

More insight into this fundamental question followed the development of DMFT methods (Georges et al., 1996), which extended the BR theory to include incoherent (inelastic) processes at finite temperature. According to this formulation, the Mott transition does assume first-order character with the associated metal–insulator coexistence dome. However, this is found only below the very low critical end-point temperature $T_c \approx 0.02 T_F$, and a smooth crossover behavior arises in the very broad intermediate-temperature interval $T_c < T < T_F$. Precisely the same features for the Mott transition

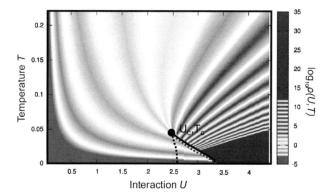

Figure 1.19 Color-coded plot of resistivity across the DMFT phase diagram of a half-filled Hubbard model. Reprinted (figure) with permission from [Vučičević, J., Terletska, H., Tanasković, D., and Dobrosavljević, V. (2013). *Phys. Rev. B* **88**, p. 075143.] Copyright (2013) by the American Physical Society.

phase diagram are found, not only in many TMO materials, but also in Mott organics we discussed above.

A visually striking illustration of what happens in the inter-mediate-temperature range is seen by color-coding the resistivity across the DMFT phase diagram (Vučičević et al., 2013), where each white band indicates another order of magnitude in resis-tivity. If one ignores the low-temperature coexistence region, one immediately notices a fan-like shape of the constant resistivity lines, characteristic of what is generally expected (Sachdev, 2011) for quantum criticality. Given the fact that the coexistence dome is a very small energy feature, one should consider the Mott transition as having weakly first-order (WFO) character. At WFO points, which are well known in standard critical phenomena (Goldenfeld, 1992), the transition assumes a first-order character only very close to the critical point; further away, the behavior is precisely that of a conventional second-order phase transition, including all aspects of the scaling phenomenology.

In the case of MITs, the QCP is expected (Dobrosavljević et al., 2012) to occur at $T = 0$. Therefore, if a QPT assumes WFO character, as we find here, we still expect to observe all features of quantum criticality in a broad intermediate-temperature region above the

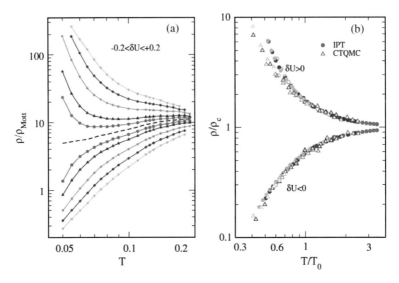

Figure 1.20 Scaling behavior of the resistivity curves in the quantum-critical region of the Mott transition at half filling, as obtained from DMFT (Terletska et al., 2011). Note the remarkable mirror symmetry of the corresponding scaling function, as experimentally found both in 2DEG systems near 2D-MIT (Simonian et al., 1997a) and also in very recent experiments in Mott organics (Furukawa et al., 2015). Reprinted (figure) with permission from [Terletska, H., Vučičević, J., Tanasković, D., and Dobrosavljević, V. (2011). *Phys. Rev. Lett.* **107**, p. 026401.] Copyright (2011) by the American Physical Society.

coexistence dome. Similar situations are not uncommon in general QC phenomena (Sachdev, 2011), since the immediate vicinity of many QCPs is often masked by the dome of an appropriate ordered state induced by critical fluctuations.

If these ideas are correct, then an appropriate scaling analysis should be able to reveal the expected QC scaling of the resistivity curves in a broad intermediate-temperature range $T_c < T < T_F$. To perform such an analysis, one needs to follow a judiciously chosen trajectory across the phase diagram, corresponding to the center of the QC region. Such a trajectory—the so-called Widom line— was identified a long time ago (Widom, 1965) in the context of conventional (thermal) critical phenomena. Directly applying these standard scaling procedures to the DMFT description of the Mott transition was first carried out a few years ago (Terletska et al., 2011), revealing compelling evidence of QC behavior. Follow-up work further established that such QC behavior is a robust feature of the Mott point both at half filling (Vučičević et al., 2013) and for doping-driven Mott transitions (Vučičević et al., 2015). The latter result also provided important new insight into the origin of the so-called "bad metal" (linear resistivity) behavior in a doped Mott insulator, a long-time puzzle (Emery and Kivelson, 1995) in the field of correlated electrons.

The discovery of the Mott QC regime within DMFT (Terletska et al., 2011) was directly motivated by the pioneering works on 2D-MIT (Kravchenko et al., 1995; Popović et al., 1997), which first revealed intriguing features of QC scaling near the MIT. The results obtained from DMFT presented surprising similarity to the experiment, both in identifying previously overlooked scaling behavior in interaction-driven MITs without disorder and in producing microscopic underpinning for early scaling phenomenology (Dobrosavljević et al., 1997) for the mirror symmetry. Still, this model calculation, based on the single-band Hubbard model at half filling, should really not be viewed as an accurate description of dilute 2DEG systems, where the experiments have been performed.

In contrast, organic Mott systems of the κ-family should be considered (Powell and McKenzie, 2011) as a much more faithful realization of the half-filled Hubbard model. This notion stimulated further experimental investigation, following the 2011 theoretical discovery, focusing on the previously overlooked intermediate-temperature region $T_c < T < T_F$. The experiments took several years of very careful work on several different materials in this family, but the results published by 2015 provided (Furukawa et al., 2015) spectacular confirmation of the DMFT theoretical predictions.

1.4.6.1 Is Wigner crystallization a Mott transition in disguise?

The original ideas of Mott (Mott, 1949), who thought about doped semiconductors, envisioned electrons hopping between well-localized atomic orbitals corresponding to donor ions. In other Mott systems, such as TMOs, the electrons travel between the atomic orbitals of the appropriate transition metal ions. In all these cases, the Coulomb repulsion restricts the occupation of such localized orbitals, leading to the Mott insulating state, but it does not provide the essential mechanism for the formation of such tightly bound electronic states. The atomic orbitals in all these examples result from the (partially screened) ionic potential within the crystal lattice.

The situation is more interesting if one considers an idealized situation describing an interacting electron gas in the absence of any periodic (or random) lattice potential due to ions. Such a physical situation is achieved, for example, when dilute carriers are injected in a semiconductor quantum well (Ando et al., 1982), where all the effects of the crystal lattice can be treated within the effective mass approximation (Ashcroft and Mermin, 1976). This picture is valid if the Fermi wavelength of the electron is much longer then the lattice spacing and the quantum mechanical dynamics of the Bloch electron can be reduced to that of a free itinerant particle with a band mass m_b. In such situations, the only potential energy in the problem corresponds to the Coulomb repulsion E_C between the electrons, which is the dominant energy scale in low-carrier-density systems. At the lowest densities, $E_C \gg E_F$, and the electrons form a Wigner crystal lattice (Wigner, 1934) to minimize the Coulomb repulsion.

Here, each electron is confined not by an ionic potential but by the formation of a deep potential well produced by repulsion from other electrons. The same mechanism prevents double occupation of such localized orbitals, and each electron in the Wigner lattice reduces to a localized $S = 1/2$ localized magnetic moment. A Wigner crystal is therefore nothing but a magnetic insulator: a Mott insulator in disguise. At higher densities, the Fermi energy becomes sufficiently large to overcome the Coulomb repulsion, and the Wigner lattice melts (Tanatar and Ceperley, 1989). The electrons then form an FL. The quantum melting of a Wigner crystal is therefore a metal–insulator transition, perhaps in many ways similar to a conventional Mott transition. What kind of phase transition is this? Despite years of effort, this important question is still not fully resolved.

What degrees of freedom play the leading role in destabilizing the Wigner crystal as it melts? Even in absence of an accepted and detailed theoretical picture describing this transition, we may immediately identify two possible classes of elementary excitations which potentially contribute to melting, as follows:

(1) *Collective charge excitations* (elastic deformations) of the Wigner crystal. In the quantum limit, these excitations have a bosonic character, but they persist and play an important role even in the semiclassical ($k_B T \gg E_F$) limit, where they contribute to the thermal melting of the Wigner lattice (Thouless, 1978). They clearly dominate in the quantum Hall regime (Chen et al., 2006), where both the spin degrees of freedom and the kinetic energy are suppressed due to Landau quantization. But for 2D-MIT in a zero magnetic field, these degrees of freedom may not be so important.

(2) *Single-particle excitations* leading to vacancy interstitial pair formation (Fig. 1.21). These excitations have a fermionic character, where the spin degrees of freedom play an important role. Recent quantum Monte Carlo simulations indicate (Cândido et al., 2001) that the effective gap for vacancy interstitial pair formation seems to collapse precisely around the quantum melting of the Wigner crystal. If these excitation dominate, then quantum melting of the Wigner crystal is a process very similar to the Mott MIT and may be expected to produce a strongly correlated FL on the metallic side. This physical picture is the central idea of this article.

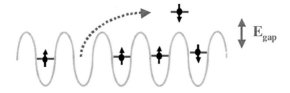

Figure 1.21 In a Wigner crystal, each electron is confined to a potential well produced by Coulomb repulsion from neighboring electrons, forming a spin-1/2 local moment. The lowest-energy particle hole excitation creates a vacancy interstitial pair (Cândido et al., 2001), which costs an energy E_{gap} comparable to the Coulomb repulsion.

We should mention, however, that the Wigner crystal melting in a zero magnetic field is believed (Tanatar and Ceperley, 1989) to be a WFO phase transition. Conventional (e.g., liquid–gas or liquid–crystal) first-order transitions are normally associated with a density discontinuity and global phase separation within the coexistence dome. For charged systems, however, global phase separation is precluded by charge neutrality (Gor'kov and Sokol, 1987). In this case, one may expect the emergence of various modulated intermediate phases, leading to bubble or stripe (Jamei et al., 2005) or possibly even stripe glass (Schmalian and Wolynes, 2000; Mahmoudian et al., 2015) order. While convincing evidence for the relevance of such nanoscale phase separation has been identified (Terletska and Dobrosavljević, 2011) in certain systems (Jaroszyński et al., 2007), recent work seems to indicate (Waintal, 2006; Clark et al., 2009) that such effects may be negligibly small for Wigner crystal melting.

A complementary line of work, where the Mott character of the transition is the focus (Pankov and Dobrosavljevic, 2008), has been the subject of recent model calculations based on DMFT approaches. These works considered a toy lattice model for the Wigner–Mott transition by examining an extended Hubbard model at quarter

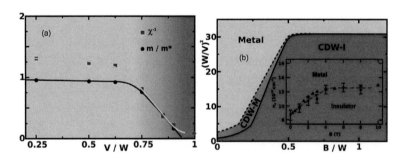

Figure 1.22 Toy model for Wigner crystalization. Results from the DMFT solution of a Hubbard model at quarter filling (Camjayi et al., 2008) show that (a) both the spin susceptibility χ and the effective mass m^* are strongly enhanced as the transition is approached. (b) Phase diagram in the presence of (parallel) magnetic field. A field-driven transition is found only sufficiently close to the insulating state in agreement with experiments. The inset shows the experimental phase diagram (Shashkin et al., 2001; Eng et al., 2002).

filling (Camjayi et al., 2008; Amaricci et al., 2010), which has also been discussed (Fratini and Merino, 2009; Merino et al., 2013) in the context of the θ-family of Mott organics. In general, a spatially uniform phase cannot become a Mott insulator away from half filling, even for arbitrarily large values of the on-site repulsion U. However, when the intersite interaction V is sufficiently large (as compared to the bandwidth W), the electronic system undergoes charge ordering, where one sublattice becomes close to half filling, while the other one become nearly empty. Such charge ordering, which is the lattice analogue of Wigner crystallization, results in a Mott insulating state, having localized spin-1/2 magnetic moments on each site of the carrier-rich sublattice.

The corresponding MIT has a character very similar to the conventional Mott transition, and the approach from the metallic side resembles the familiar BR scenario (Brinkman and Rice, 1970). One finds strong enhancement in both the spin susceptibility and the QP effective mass m^* in agreement with experiments. A new feature, which differs from the standard Mott transition at half filling, is the response to (parallel) magnetic fields (Zeeman coupling). This calculation finds (Camjayi et al., 2008), in agreement with experiment on 2DEG systems (Shashkin et al., 2001; Eng et al., 2002), that a field-driven MIT arises only sufficiently close to the $B = 0$ MIT, while deeper in the metallic phase even full spin polarization cannot completely destroy the metal.

1.5 Conclusions

In this chapter we presented evidence suggesting that 2D-MIT found in ultraclean 2DEG devices should be viewed as an interaction-driven MIT with many features in common with conventional Mott systems. Striking similarities were established between thermodynamic and transport properties of the respective experiments but also with the predictions of generic theoretical models describing Mott and Wigner–Mott transitions. The resulting DMFT physical picture seems to offer, for quantum fluids, a perspective comparable to what the very successful van der Waals theory provided for the classical liquid–gas critical point. It should give the proper starting point for future theoretical studies of more complicated experiments containing nontrivial interplay between strong correlations and disorder, as described in the chapter by D. Popovic. This fascinating research direction remains a challenge for upcoming theoretical work.

Acknowledgments

The authors are indebted to their collaborators in the field: E. Abrahams, S. Chakravarty, S. Florens, S. Fratini, K. Haule, G. Kotliar, E. Manousakis, E. Miranda, G. Moeller, S. Pankov, D. Popović, Y. Pramudya, M. Radonjić, A. Ralko, M. J. Rozenberg, A. E. Ruckenstein, H. Terletska, and J. Vučičević. The authors have benefitted from very useful discussions with A. Georges, A. M. Finkel'stein, S. Hartnoll, K. Kanoda, S. Kravchenko, R. H. McKenzie, Z. Y. Meng, A. J. Millis, P. Monceau, A. Punnoose, M. P. Sarachik, J. Saunders, T. Senthil, J. Schmalian, G. Sordi, A. M. Tramblay, M. Vojta, and J. Zaanen. V. D. was supported by the NSF grant DMR-1410132 and the National High Magnetic Field Laboratory. D. T. acknowledges support from the Serbian Ministry of Education, Science and Technological Development under Project No. ON171017.

References

Abrahams, E., Anderson, P. W., Licciardello, D. C., and Ramakrishnan, T. V. (1979). *Phys. Rev. Lett.* **42**, p. 673.

Abrahams, E., Kravchenko, S. V., and Sarachik, M. P. (2001). *Rev. Mod. Phys.* **73**, p. 251.

Amaricci, A., Camjayi, A., Haule, K., Kotliar, G., Tanasković, D., and Dobrosavljević, V. (2010). *Phys. Rev. B* **82**, p. 155102.

Altshuler, B. L., Maslov, D. L., and Pudalov, V. M. (2001). *Physica (Amsterdam)* **9E**, p. 209.

Anderson, P. (1958). *Phys. Rev.* **109**, p. 1492.

Anderson, P. W. (1959). *Phys. Rev.* **115**, p. 2.

Ando, T., Fowler, A. B., and Stern, F. (1982). *Rev. Mod. Phys.* **54**, p. 437.

Andrade, E. C., Miranda, E., and Dobrosavljević, V. (2009). *Phys. Rev. Lett.* **102**, p. 206403.

Ashcroft, N. W., and Mermin, D. (1976). *Solid State Physics* (Saunders College, Philadelphia).

Brinkman, W. F., and Rice, T. (1970). *Phys. Rev. B* **2**, p. 4302.

Camjayi, A., Haule, K., Dobrosavljević, V., and Kotliar, G. (2008). *Nat. Phys.* **4**, p. 932.

Cândido, L., Bernu, B., and Ceperley, D. M. (2004). *Phys. Rev. B* **70**, p. 094413.

Cândido, L., Phillips, P., and Ceperley, D. M. (2001). *Phys. Rev. Lett.* **86**, p. 492.

Casey, A., Patel, H., Nyèki, J., Cowan, B. P., and Saunders, J. (2003). *Phys. Rev. Lett.* **90**, p. 115301.

Chen, Y. P., Sambandamurthy, G., Wang, Z. H., Lewis, R. M., Engel, L. W., Tsui, D. C., Ye, P. D., Pfeiffer, L. N., and West, K. W. (2006). *Nat. Phys.* **2**, p. 452.

Dobbs, B. J. T. (2002). *The Janus Faces of Genius: The Role of Alchemy in Newtons Thought* (Cambridge University Press, UK).

Clark, B. K., Casula, M., and Ceperley, D. M. (2009). *Phys. Rev. Lett.* **103**, p. 055701.

Dobrosavljević, V., Abrahams, E., Miranda, E., and Chakravarty, S. (1997). *Phys. Rev. Lett.* **79**, p. 455.

Dobrosavljević, V., Trivedi, N., and Valles Jr., J. M. (2012). *Conductor Insulator Quantum Phase Transitions* (Oxford University Press, UK).

Emery, V. J., and Kivelson, S. A. (1995). *Phys. Rev. Lett.* **74**, p. 3253.

Elsässer, S., Wu, D., Dressel, M., and Schlueter, J. A. (2012). *Phys. Rev. B* **86**, p. 155150.

Eng, K., Feng, X. G., Popović, D., and Washburn, S. (2002). *Phys. Rev. Lett.* **88**, p. 136402.

Finkel'stein, A. M. *Zh. Eksp. Teor. Fiz.* **84**, p. 168; (1983) *Sov. Phys. JETP* **57**, p. 97.

Finkel'stein, A. M. (1984). *Zh. Eksp. Teor. Fiz.* **86**, p. 367; (1983) *Sov. Phys. JETP* **59**, p. 212.

Flouquet, J., Lasjaunias, J. C., Peyrard, J., and Ribault, M. (1982). *J. Appl. Phys.* **53**, pp. 2127–2130.

Flouquet, J. (2005). *Progress in Low Temperature Physics*, Vol. 15 (Elsevier, Amsterdam) pp. 139–281.

Fowler, A. B., Fang, F. F., Howard, W. E., and Stiles, P. J. (1966). *Phys. Rev. Lett.* **16**, p. 901.

Fratini, S., and Merino, J. (2009). *Phys. Rev. B* **80**, p. 165110.

Furukawa, T., Miyagawa, K., Taniguchi, H., Kato, R., and Kanoda, K. (2015). *Nat. Phys.* **11**, pp. 221–224.

Georges, A., Kotliar, G., Krauth, W., and Rozenberg, M. J. (1996). *Rev. Mod. Phys.* **68**, p. 13.

Goldenfeld, N. (1992). *Lectures on Phase Transitions and the Renormalization Group* (Addison-Wesley, Reading).

Goodenough, J. B. (1963). *Magnetism and the Chemical Bond* (John Wiley & Sons, New York- London).

Gor'kov, L. P., and Sokol, A. V. (1987). *JETP Lett.* **46**, p. 420.

Gull, E., Millis, A. J., Lichtenstein, A. I., Rubtsov, A. N., Troyer, M., and Werner, P. (2011). *Rev. Mod. Phys.* **83**, p. 349.

Hewson, A. C. (1993). *The Kondo Problem to Heavy Fermions* (Cambrige University Press, Cambridge).

Hubbard, J. (1963). *Proc. R. Soc. (London) A* **276**, p. 238.

Jacko, A. C., Fjaerestad, J. O., and Powell, B. J. (2009). *Nat. Phys.* **5**, pp. 422–425.

Jamei, R., Kivelson, S., and Spivak, B. (2005). *Phys. Rev. Lett.* **94**, p. 056805.

Jaroszyński, J., Andrearczyk, T., Karczewski, G., Wr?bel, J., Wojtowicz, T., Popović, D., and Dietl, T. (2007). *Phys. Rev. B* **76**, p. 045322.

Kotliar, G., and Ruckenstein, A. E. (1986). *Phys. Rev. Lett.* **57**, p. 1362.

Kotliar, G., Savrasov, S. Y., Haule, K., Oudovenko, V. S., Parcollet, O., and Marianetti, C. A. (2006). *Rev. Mod. Phys.* **78**, p. 865.

Kravchenko, S. V., and Sarachik, M. P. (2004). *Rep. Prog. Phys.* **67**, p. 1.

Kravchenko, S. V., Mason, W. E., Bowker, G. E., Furneaux, J. E., Pudalov, V. M., and D'Iorio, M. (1995). *Phys. Rev. B* **51**, p. 7038.

Kurosaki, Y., Shimizu, Y., Miyagawa, K., Kanoda, K., and Saito, G. (2005). *Phys. Rev. Lett.* **95**, p. 177001.

Landau, L. D. (1957). *Sov. Phys. JETP* **3**, p. 920.

Lee, P. A., and Ramakrishnan, T. V. (1985). *Rev. Mod. Phys.* **57**, p. 287.

Limelette, P., Wzietek, P., Florens, S., Georges, A., Costi, T. A., Pasquier, C., Jerome, D., Meziere, C., and Batail, P. (2003). *Phys. Rev. Lett.* **91**, p. 016401.

Mahmoudian, S., Rademaker, L., Ralko, A., Fratini, S., and Dobrosavljević, V. (2015). *Phys. Rev. Lett.* **115**, p. 025701.

Merino, J., Ralko, A., and Fratini, S. (2013). *Phys. Rev. Lett.* **111**, p. 126403.

Millis, A. J. (1993). *Phys. Rev. B* **48**, p. 7183.

Misguich, G., Bernu, B., Lhuillier, C., and Waldtmann, C. (1998). *Phys. Rev. Lett.* **81**, p. 1098.

Miranda, E., and Dobrosavljević, V. (2005). *Rep. Prog. Phys.* **68**, p. 2337.

Moeller, G., Dobrosavljević, V., and Ruckenstein, A. E. (1999). *Phys. Rev. B* **59**, p. 6846.

Mokashi, A., Li, S., Wen, B., Kravchenko, S. V., Shashkin, A. A., Dolgopolov, V. T., and Sarachik, M. P. (2012). *Phys. Rev. Lett.* **109**, p. 096405.

Mott, N. F. (1949). *Proc. Phys. Soc. (London) A* **62**, p. 416.

Mott, N. F. (1990). *Metal-Insulator Transition* (Taylor & Francis, London).

Pankov, S., and Dobrosavljević, V. (2008). *Physica B* **403**, p. 1440.

Park, H., Haule, K., and Kotliar, G. (2008). *Phys. Rev. Lett.* **101**, p. 186403.

Pfeiffer, L. N., and West, K. W. (2006). *Nat. Phys.* **2**, p. 452.

Pines, D., and Nozières, P. (1965). *The Theory of Quantum Liquids* (Benjamin, New York).

Popović, D., Fowler, A. B., and Washburn, S. (1997). *Phys. Rev. Lett.* **79**, p. 1543.

Powell, B. J. and McKenzie, R. H. (2011). *Rep. Prog. Phys.* **74**, p. 056501.

Prus, O., Yaish, Y., Reznikov, M., Sivan, U., and Pudalov, V. (2003). *Phys. Rev. B* **67**, p. 205407.

Pudalov, V., Brunthaler, G., Prinz, A., and Bauer, G. (1998). *Physica E* **3**, pp. 79–88.

Punnoose, A., and Finkel'stein, A. M. (2001). *Phys. Rev. Lett.* **88**, p. 016802.

Radonjić, M. M., Tanaskovic, D., Dobrosavljevic, V., and Haule, K. (2010). *Phys. Rev. B* **81**, p. 075118.

Radonjić, M. M., Tanasković, D., Dobrosavljević, V., Haule, K., and Kotliar, G. (2012). *Phys. Rev. B* **85**, p. 085133.

Sachdev, S. (2011). *Quantum Phase Transitions,* 2nd Edition (Cambridge University Press, UK).

Sarachik, M. P. (1995). In *Metal-Insulator Transitions Revisited*, eds. Edwards, P., and Rao, C. N. R. (Taylor and Francis, London).

Schmalian, J., and Wolynes, P. G. (2000). *Phys. Rev. Lett.* **85**, p. 836.

Shashkin, A. A., Kravchenko, S. V., Dolgopolov, V. T., and Klapwijk, T. M. (2001). *Phys. Rev. Lett.* **87**, p. 086801.

Shashkin, A. A., Kravchenko, S. V., Dolgopolov, V. T., and Klapwijk, T. M. (2002). *Phys. Rev. B* **66**, p. 073303.

Simonian, D., Kravchenko, S. V., and Sarachik, M. P. (1997a). *Phys. Rev. B* **55**, p. R13421.

Simonian, D., Kravchenko, S. V., Sarachik, M. P., and Pudalov, V. M. (1997b). *Phys. Rev. Lett.* **79**, p. 2304.

Shklovskii, B. I., and Efros, A. L. (1984). *Electronic Properties of Doped Semiconductors* (Springer-Verlag, Berlin, Heidelberg).

Slater, J. C. (1934). *Rev. Mod. Phys.* **6**, p. 209.

Slater, J. C. (1951). *Phys. Rev.* **82**, p. 538.

Stewart, G. R. (1984). *Rev. Mod. Phys.* **56**, p. 755.

Tanasković, D., Haule, K., Kotliar, G., and Dobrosavljević, V. (2011). *Phys. Rev. B* **84**, p. 115105.

Tanatar, B., and Ceperley, D. M. (1989). *Phys. Rev. B* **39**, p. 5005.

Terletska, H., and Dobrosavljević, V. (2011). *Phys. Rev. Lett.* **106**, p. 186402.

Terletska, H., Vučičević, J., Tanasković, D., and Dobrosavljević, V. (2011). *Phys. Rev. Lett.* **107**, p. 026401.

Thouless, D. J. (1978). *J. Phys. C: Solid State Phys.* **11**, p. L189.

Ting, C. S., Lee, T. K., and Quinn, J. J. (1975). *Phys. Rev. Lett.* **34**, p. 870.

Vollhardt, D. (1984). *Rev. Mod. Phys.* **56**, p. 99.

Vučičević, J., Terletska, H., Tanasković, D., and Dobrosavljević, V. (2013). *Phys. Rev. B* **88**, p. 075143.

Vučičević, J., Tanasković, D., Rozenberg, M. J., and Dobrosavljević, V. (2015). *Phys. Rev. Lett.* **114**, p. 246402.

Waintal, X. (2006). *Phys. Rev. B* **73**, p. 075417.

Widom, B. (1965). *J. Chem. Phys.* **43**, p. 3898.

Wigner, E. (1934). *Phys. Rev.* **46**, p. 1002.

Zala, G., Narozhny, B. N., and Aleiner, I. L. (2001). *Phys. Rev. B* **64**, p. 214204.

Zhang, Y., and Das Sarma, S. (2005). *Phys. Rev. B* **71**, p. 045322.

Chapter 2

Metal–Insulator Transition in a Strongly Correlated Two-Dimensional Electron System

A. A. Shashkin[a] and S. V. Kravchenko[b]

[a]*Institute of Solid State Physics, Chernogolovka, Moscow District 142432, Russia*
[b]*Physics Department, Northeastern University, Boston, Massachusetts 02115, USA*
shashkin@issp.ac.ru, s.kravchenko@northeastern.edu

Experimental results on the metal–insulator transition and related phenomena in strongly interacting two-dimensional electron systems are discussed. Special attention is given to recent results for the strongly enhanced spin susceptibility, effective mass, and thermopower in low-disordered silicon MOSFETs.

2.1 Strongly and Weakly Interacting 2D Electron Systems

In two-dimensional (2D) electron systems, the electrons are confined in a 1D potential well and are free to move in a plane. The strongly interacting limit is reached in such systems at low

Strongly Correlated Electrons in Two Dimensions
Edited by Sergey Kravchenko
Copyright © 2017 Pan Stanford Publishing Pte. Ltd.
ISBN 978-981-4745-37-6 (Hardcover), 978-981-4745-38-3 (eBook)
www.panstanford.com

electron densities where the kinetic energy is overwhelmed by the energy of the electron–electron interactions. The interaction strength is characterized by the ratio between the Coulomb energy and the Fermi energy, $r_s^* = E_{ee}/E_F$. If we assume that the effective electron mass is equal to the band mass, the interaction parameter r_s^* in the single-valley case reduces to the Wigner–Seitz radius,

$$r_s = 1/(\pi n_s)^{1/2} a_B, \qquad (2.1)$$

and, therefore, increases as the electron density, n_s, decreases (here a_B is the Bohr radius in semiconductor). Candidates for the ground state of the system include a Wigner crystal characterized by spatial and spin ordering (Wigner, 1934), a ferromagnetic Fermi liquid with spontaneous spin ordering (Stoner, 1946), paramagnetic Fermi liquid (Landau, 1957), etc. In the strongly interacting limit ($r_s \gg 1$), no analytical theory has been developed so far. Numeric simulations (Tanatar and Ceperley, 1989) show that the Wigner crystallization is expected in a very dilute regime, when r_s reaches approximately 35. More recent refined numeric simulations (Attaccalite et al., 2002) have predicted that prior to the Wigner crystallization, in the range of the interaction parameter $25 \leq r_s \leq 35$, the ground state of the system is a strongly correlated ferromagnetic Fermi liquid. At higher electron densities, $r_s \approx 1$ (weakly interacting regime), the electron liquid is expected to be paramagnetic, with the effective mass, m, and Landé g-factor renormalized by interactions. In addition to the ferromagnetic Fermi liquid, other intermediate phases between the Wigner crystal and the paramagnetic Fermi liquid may also exist.

In real 2D electron systems, the always-present disorder leads to a drastic change of the above simplified picture, which dramatically complicates the problem. According to the scaling theory of localization (Abrahams et al., 1979), in an infinite disordered noninteracting 2D system, all electrons become localized at zero temperature and in a zero magnetic field. At finite temperatures, regimes of strong and weak localizations can be distinguished:

(i) if the conductivity of the 2D electron layer has an activated character, the resistivity diverges exponentially as $T \to 0$; and

(ii) in the opposite limit (the so-called weak localization), the resistivity increases logarithmically with decreasing temperature—an effect originating from the increased probability of electron backscattering from impurities to the starting point. The incorporation of weak interactions ($r_s < 1$) between the electrons adds to the localization (Altshuler et al., 1980). However, for weak disorder and $r_s \geq 1$, a possible metallic ground state was predicted (Castellani et al., 1984; Finkelstein, 1983, 1984).

In view of the competition between the interactions and disorder, one can consider high- and low-disorder limits. In highly disordered electron systems, the range of low densities is not accessible as the strong (Anderson) localization sets in. This corresponds to the weakly interacting limit in which an insulating ground state is expected. Much more interesting is the case of low-disorder electron systems because low electron densities corresponding to the strongly interacting limit become accessible. According to the renormalization group analysis for multivalley 2D systems (Punnoose and Finkelstein, 2005), the strong electron–electron interactions can stabilize the metallic ground state, leading to the existence of a metal–insulator transition in a zero magnetic field.

In this chapter, we will focus on experimental results obtained in low-disordered strongly interacting 2D electron systems; in particular, in (100)-silicon metal-oxide-semiconductor field-effect transistors (MOSFETs). Due to the relatively large effective mass, relatively small dielectric constant, and the presence of two valleys in the spectrum, the interaction parameter in silicon MOSFETs is an order of magnitude larger at the same electron densities compared to that in the 2D electron system in n-GaAs/AlGaAs heterostructures: except at extremely low electron densities, the latter electron system can be considered weakly interacting. Indeed, the observed effects of strong electron–electron interactions are more pronounced in silicon MOSFETs compared to n-GaAs/AlGaAs heterostructures, although the fractional quantum Hall effect, attributed to electron–electron interactions, has not been reliably established in silicon MOSFETs.

2.2 Zero-Field Metal–Insulator Transition

As we have already mentioned, no conducting state was expected in a zero magnetic field, at least in weakly interacting 2D electron systems. Therefore, it came as a surprise when the behavior consistent with the existence of a metallic state and the metal–insulator transition was found in strongly interacting 2D electron systems in silicon (Kravchenko et al., 1994, 1995; Zavaritskaya and Zavaritskaya, 1987).

Provided that the temperature dependences of the resistance are strong, the curves with positive (negative) derivative $d\rho/dT$ are indicative of a metal (insulator) states (Abrahams et al., 2001; Kravchenko and Sarachik, 2004; Sarachik and Kravchenko, 1999). If extrapolation of $\rho(T)$ to $T = 0$ is valid, the critical point for the metal–insulator transition is given by $d\rho/dT = 0$. In a low-disordered 2D electron system in silicon MOSFETs, the resistivity at a certain "critical" electron density shows virtually no temperature dependence over a wide range of temperatures (Kravchenko and Klapwijk, 2000; Sarachik and Kravchenko, 1999) (Fig. 2.1a,b). This curve separates those with positive and negative $d\rho/dT$ nearly symmetrically at temperatures above 0.2 K (Abrahams et al., 2001; Kravchenko and Sarachik, 2004). Assuming that it remains flat down to $T = 0$, one obtains the critical point which corresponds to a resistivity $\rho \approx 3h/e^2$.

To verify whether or not the separatrix corresponds to the critical density, an independent determination of the critical point is necessary: comparison of values obtained using different criteria provides an experimental test of whether or not a true MIT exists at $B = 0$. One criterion (the "derivative criterion") was described above; its weakness is that it requires extrapolation to zero temperature. A second criterion can be applied based on an analysis of a temperature-independent characteristic, namely, the localization length L extrapolated from the insulating phase. These two methods have been applied to low-disordered silicon MOSFETs by Shashkin et al. (Shashkin et al., 2001a) and Jaroszynski et al. (Jaroszynski et al., 2002).

Deep in the insulating phase, the temperature dependence of the resistance obeys the Efros–Shklovskii variable-range hopping form

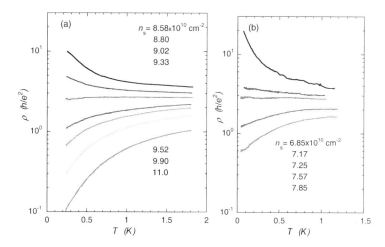

Figure 2.1 Resistivity versus temperature in two Si-MOSFET samples from different sources. (a) High-mobility sample provided by V. M. Pudalov. Reprinted from Sarachik, M. P. and Kravchenko, S. V. (1999). *Proc. Natl. Acad. Sci. USA* **96**, pp. 5900–5902. Copyright (1999) National Academy of Sciences. (b) sample fabricated by R. Heemskerk and T. M. Klapwijk. Reprinted (figure) with permission from [Kravchenko, S. V. and Klapwijk, T. M. (2000). *Phys. Rev. Lett.* **84**, pp. 2909–2912.] Copyright (2000) by the American Physical Society.

(Mason et al., 1995); however, closer to the critical electron density and at temperatures that are not too low, the resistance has an activated form $\rho \propto e^{E_a/k_B T}$ (Adkins et al., 1976; Dolgopolov et al., 1992; Pudalov et al., 1993; Shashkin et al., 1994a, 1994b) due to thermal activation to the mobility edge. Figure 2.2 shows the activation energy E_a as a function of the electron density (diamonds); the data can be approximated by a linear function which yields, within the experimental uncertainty, the same critical electron density as the "derivative criterion" ($n_c \approx 0.795 \times 10^{11}$ cm^{-2} for the sample of Fig. 2.2).

The critical density can also be determined by studying the nonlinear current–voltage I–V characteristics on the insulating side of the transition. A typical low-temperature I–V curve is close to a step-like function: the voltage rises abruptly at low current and then saturates, as shown in the inset to Fig. 2.2; the magnitude of the step is 2 V_c. The curve becomes less sharp at higher temperatures, yet

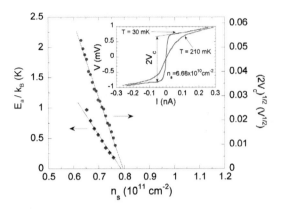

Figure 2.2 Activation energy (diamonds) and square root of the threshold voltage (circles) versus electron density in a zero magnetic field in a low-disordered silicon MOSFET. The inset shows current–voltage characteristics recorded at ~30 and 210 mK, as labeled; note that the threshold voltage is essentially independent of temperature. Reprinted (figure) with permission from [Shashkin, A. A., Kravchenko, S. V., and Klapwijk, T. M. (2001a). *Phys. Rev. Lett.* **87**, p. 266402.] Copyright (2001) by the American Physical Society.

the threshold voltage, V_c, remains essentially unchanged. Closer to the MIT, the threshold voltage decreases, and at $n_s = n_c \approx 0.795 \times 10^{11}$ cm^{-2}, the I–V curve is strictly linear (Shashkin et al., 2001a). According to Polyakov and Shklovskii (1993) and Shashkin et al. (1994a, 1994b), the breakdown of the localized phase occurs when the localized electrons at the Fermi level gain enough energy to reach the mobility edge in an electric field, V_c/d, over a distance given by the localization length, L, which is temperature independent:

$$eV_c(n_s)\, L(n_s)/d = E_a(n_s)$$

(here d is the distance between the potential probes). The dependence of $V_c^{1/2}(n_s)$ near the MIT is linear, as shown in Fig. 2.2 by closed circles, and its extrapolation to zero threshold value again yields approximately the same critical electron density as the previous criteria. The linear dependence $V_c^{1/2}(n_s)$, accompanied by linear $E_a(n_s)$, signals the localization length diverging near the critical density: $L(n_s) \propto 1/(n_c - n_s)$.

These experiments indicate that in low-disorder samples, the two methods—one based on extrapolation of $\rho(T)$ to zero temperature and a second based on the behavior of the temperature-independent localization length—yield the same critical electron density n_c. This implies that the separatrix remains "flat" (or extrapolates to a finite resistivity) at zero temperature. Since one of the methods is independent of temperature, this equivalence supports the existence of a true $T = 0$ MIT in low-disorder samples in a zero magnetic field.

Additional confirmation in favor of zero-temperature zero-field metal–insulator transition is provided by magnetic (Shashkin, 2005) and thermopower (Mokashi et al., 2012) measurements, as described in subsequent sections. We will argue that the metal–insulator transition in silicon samples with very low level of disorder is driven by interactions.

2.3 Possible Ferromagnetic Transition

In 2000, it was experimentally found that the ratio between the spin and the cyclotron splittings in silicon MOSFETs strongly increases at low electron densities (Kravchenko et al., 2000). The spin splitting is proportional to the g-factor, while the cyclotron splitting is inversely proportional to the effective mass; therefore, their ratio is determined by the product g^*m^* which is in turn proportional to the spin susceptibility (here m^* is the renormalized effective mass and g^* is the renormalized g-factor). The strong enhancement of the spin susceptibility has indicated that at low electron densities, the system behaves well beyond the weakly interacting Fermi liquid.

Application of a magnetic field parallel to the 2D plane promotes a strong positive magnetoresistance which, however, saturates at a certain density-dependent value of the field, B_c. This saturation has been shown to correspond to the onset of the full spin polarization (Okamoto et al., 1999; Vitkalov et al., 2000); therefore, the product g^*m^* can be recalculated from the parallel-field magnetoresistance data. Shashkin et al. (Shashkin et al., 2001b) scaled the magnetoresistivity in the spirit of the theory (Dolgopolov and Gold, 2000) which predicted that at $T = 0$, the normalized magnetoresistance is a universal function of the degree of spin polarization,

$$P \equiv g^* \mu_B B_\| / 2 E_F = g^* m^* \mu_B B_\| / \pi \hbar^2 n_s \tag{2.2}$$

(here, the twofold valley degeneracy in silicon has been taken into account). Shashkin et al. (Shashkin et al., 2001b) scaled the data obtained in the limit of very low temperatures where the magnetoresistance becomes temperature independent and, therefore, can be considered to be at its $T = 0$ value. In this regime, the normalized magnetoresistance, $\rho(B_\|)/\rho(0)$, measured at different electron densities, collapses onto a single curve when plotted as a function of $B_\|/B_c$ (here B_c is the scaling parameter, normalized to correspond to the magnetic field B_{sat} at which the magnetoresistance saturates). An example of how $\rho(B_\|)$, plotted in Fig. 2.3a, can be scaled onto a universal curve is shown in Fig. 2.3b. The resulting function is described reasonably well by the theoretical dependence predicted by Dolgopolov and Gold. The quality of the scaling is remarkably good for $B_\|/B_c \leq 0.7$ in the electron density range 1.08×10^{11} to 10×10^{11} cm^{-2}. As shown in Fig. 2.4, the scaling parameter is proportional over a wide range of electron densities to the deviation of the electron density from its critical value: $B_c \propto (n_s - n_\chi)$.

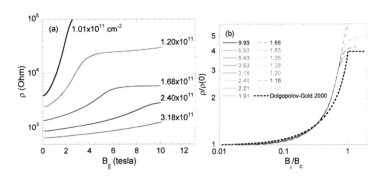

Figure 2.3 (a) Low-temperature magnetoresistance of a clean silicon MOSFET in a parallel magnetic field at different electron densities above n_c. (b) Scaled curves of the normalized magnetoresistance versus $B_\|/B_c$. The electron densities are indicated in units of 10^{11} cm^{-2}. Also shown by a thick dashed line is the normalized magnetoresistance calculated by Dolgopolov and Gold (2000). Reprinted (figure) with permission from [Shashkin, A. A., Kravchenko, S. V., Dolgopolov, V. T., and Klapwijk, T. M. (2001b). *Phys. Rev. Lett.* **87**, p. 086801.] Copyright (2001b) by the American Physical Society.

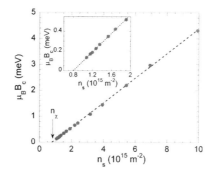

Figure 2.4 Scaling parameter B_c (corresponding to the field required for full spin polarization) as a function of the electron density. An expanded view of the region near n_χ is displayed in the inset. Reprinted (figure) with permission from [Shashkin, A. A., Kravchenko, S. V., Dolgopolov, V. T., and Klapwijk, T. M. (2001b). *Phys. Rev. Lett.* **87**, p. 086801.] Copyright (2001b) by the American Physical Society.

The fact that the parallel magnetic field required to produce complete spin polarization, $B_c \propto n_s/g^*m^*$, tends to vanish at a finite electron density $n_\chi \approx 8 \times 10^{10}$ cm^{-2} (which is close to the critical density n_c for the metal–insulator transition in this electron system) points to a sharp increase of the spin susceptibility, $\chi \propto g^*m^*$, and possible ferromagnetic instability in dilute silicon MOSFETs.

The spin susceptibility, χ, can be calculated using the above data. In the clean limit, the magnetic field required to fully polarize the electron spins is related to the g-factor and the effective mass by the equation

$$g^*\mu_B B_c = 2E_F = \pi \hbar^2 n_s/m^*.\qquad(2.3)$$

Therefore, the spin susceptibility, normalized by its "noninteracting" value, can be calculated as

$$\frac{\chi}{\chi_0} = \frac{g^*m^*}{g_0 m_b} = \frac{\pi \hbar^2 n_s}{2\mu_B B_c m_b}.\qquad(2.4)$$

The results of this recalculation are shown in Fig. 2.5. One can see that the spin susceptibility increases by a factor of 5 compared to its nonrenormalized value.

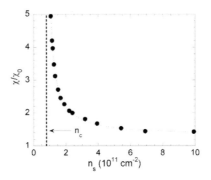

Figure 2.5 The product $g^* m^*$ versus electron density obtained from the data for B_c. The critical electron density obtained from transport measurements is indicated by the dashed line. Reprinted (figure) with permission from [Shashkin, A. A., Kravchenko, S. V., Dolgopolov, V. T., and Klapwijk, T. M. (2001b). *Phys. Rev. Lett.* **87**, p. 086801.] Copyright (2001b) by the American Physical Society.

2.4 Effective Mass or *g*-Factor?

In principle, the increase of the spin susceptibility could be due to an enhancement of either g^* or m^* (or both). Shashkin et al. (Shashkin et al., 2002) analyzed the data for the temperature dependence of the conductivity in a zero magnetic field in spirit of the theory (Zala et al., 2001). It turned out that it is the effective mass, rather than the g-factor, that sharply increases at low electron densities (Shashkin et al., 2002) (Fig. 2.6, left). It was found that the magnitude of the mass does not depend on the degree of spin polarization, indicating a spin-independent origin of the effective mass enhancement (Shashkin et al., 2003). It was also found that the relative mass enhancement is system and disorder independent and is determined by electron–electron interactions only (Shashkin et al., 2007).

In addition to transport measurements, thermodynamic measurements of the magnetocapacitance and magnetization of a 2D electron system in low-disordered silicon MOSFETs were performed, and very similar results for the spin susceptibility, effective mass, and g-factor were obtained (Anissimova et al., 2006; Khrapai et al., 2003; Shashkin et al., 2006) (Fig. 2.6, right). The Pauli

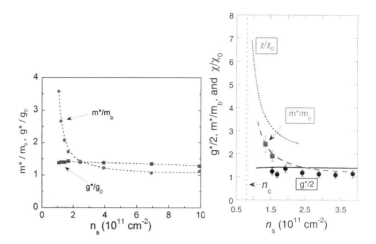

Figure 2.6 Left: The effective mass and g-factor versus electron density determined from an analysis of the temperature-dependent conductivity and parallel-field magnetoresistance. The dashed lines are guides to the eye. Reprinted (figure) with permission from [Shashkin, A. A., Kravchenko, S. V., Dolgopolov, V. T., and Klapwijk, T. M. (2002). *Phys. Rev. B* **66**, p. 073303.] Copyright (2002) by the American Physical Society. Right: The effective mass (squares) and g-factor (circles) as a function of the electron density obtained by magnetization measurements in perpendicular and tilted magnetic fields. The solid and long-dashed lines represent, respectively, the g-factor and effective mass, previously obtained from transport measurements (Shashkin et al., 2002), and the dotted line is the Pauli spin susceptibility obtained by magnetization measurements in parallel magnetic fields (Shashkin et al., 2006). The critical density n_c for the metal–insulator transition is indicated. Reprinted (figure) with permission from [Anissimova, S., Venkatesan, A., Shashkin, A. A., Sakr, M. R., Kravchenko, S. V., and Klapwijk, T. M. (2006). *Phys. Rev. Lett.* **96**, p. 046409.] Copyright (2006) by the American Physical Society.

spin susceptibility behaves critically close to the critical density n_c for the $B = 0$ metal–insulator transition: $\chi \propto n_s/(n_s - n_\chi)$. This is in favor of the occurrence of a spontaneous spin polarization (either Wigner crystal or ferromagnetic liquid) at low n_s, although in currently available samples, the residual disorder conceals the origin of the low-density phase. The effective mass increases sharply with decreasing density while the enhancement of the g-factor is weak and practically independent of n_s. Unlike in the Stoner scenario, it is the effective mass that is responsible for the dramatically enhanced spin susceptibility at low electron densities.

Yet another way to estimate the behavior of the effective mass is to measure the thermoelectric power. The thermopower is defined as the ratio of the thermoelectric voltage to the temperature difference, $S = -\Delta V/\Delta T$. Based on Fermi liquid theory, Dolgopolov and Gold (Dolgopolov and Gold, 2011; Gold and Dolgopolov, 2011) obtained the following expression for the diffusion thermopower of strongly interacting 2D electrons in the low-temperature regime:

$$S = -\alpha \frac{2\pi k_B^2 m^* T}{3e\hbar^2 n_s},\qquad (2.5)$$

where k_B is Boltzmann's constant. This expression, which resembles the well-known Mott relation for noninteracting electrons, was shown to hold for the strongly interacting case provided one includes the parameter α that depends on both disorder (Faniel et al., 2007; Fletcher et al., 1997; Goswami et al., 2009) and interaction strength (Dolgopolov and Gold, 2011; Gold and Dolgopolov, 2011). The dependence of α on electron density is rather weak, and the main effect of electron–electron interactions is to suppress the thermopower S.

According to Eq. 2.5, $(-S/T)$ is proportional to (m^*/n_s) and, therefore, the measurements of the thermopower yield the effective mass. The divergent behavior of the thermopower is evident when plotted as the inverse quantity, $(-1/S)$, versus electron density in Fig. 2.7 (left). Figure 2.7 (right) shows $(-T/S)$ plotted as a function of n_s. The data collapse onto a single curve demonstrating that the thermopower S is a linear function of temperature. In turn, the ratio $(-T/S)$ is a function of electron density n_s of form:

$$(-T/S) \propto (n_s - n_t)^x.\qquad (2.6)$$

Fits to this expression indicate that the thermopower diverges with decreasing electron density with a critical exponent $x = 1.0 \pm 0.1$ at the density $n_t = 7.8 \pm 0.1 \times 10^{10}$ cm^{-2} that is close to (or the same as) the density for the metal–insulator transition, $n_c \approx 8 \times 10^{10}$ cm^{-2}, obtained from resistivity measurements in this low-disorder electron system. The measured $(-T/S)$, shown in Fig. 2.7 (right), decreases linearly with decreasing electron density,

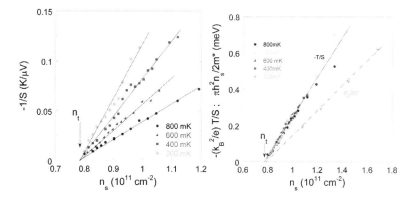

Figure 2.7 Left: The inverse thermopower as a function of electron density at different temperatures. The solid lines denote linear fits to the data and extrapolate to zero at a density n_t. Right: $(-T/S)$ versus electron density for different temperatures. The solid line is a linear fit which extrapolates to zero at n_t. Also shown is the effective mass obtained for the same samples by transport measurements (Shashkin et al., 2002). The dashed line is a linear fit. Reprinted (figure) with permission from [Mokashi, A., Li, S.,Wen, B., Kravchenko, S. V., Shashkin, A. A., Dolgopolov, V. T., and Sarachik, M. P. (2012). *Phys. Rev. Lett.* **109**, p. 096405.] Copyright (2012) by the American Physical Society.

extrapolating to zero at n_t and indicating a strong increase of the effective mass by more than an order of magnitude. The results thus imply a divergence of the electron mass at the density n_t: $m^* \propto n_s/(n_s - n_t)$— a behavior that is typical in the vicinity of an interaction-induced phase transition.

It is interesting to compare these results with the effective mass obtained earlier for the same samples. As seen in Fig. 2.7 (right), the two data sets display similar behavior. However, the thermopower data do not yield the absolute value of m^* because of uncertainty in the coefficient α in Eq. 2.5. The value of m^* can be extracted from the thermopower data by requiring that the two data sets in Fig. 2.7 (right) correspond to the same value of mass in the range of electron densities where they overlap. Determined from the ratio of the slopes, this yields a coefficient $\alpha \approx 0.18$. The corresponding mass enhancement in the critical region reaches $m^*/m_b \approx 25$ at $n_s \approx 8.2 \times 10^{10}$ cm^{-2}, where the band mass $m_b = 0.19 m_e$ and m_e is the free electron mass. The mass $m^* \approx 5 m_e$ exceeds by far the values of the effective mass obtained from previous experiments. It is important to note that the thermopower experiment includes data for electron densities that are much closer to the critical point than the earlier measurements, and reports much larger enhancement of the effective mass for reasons explained below.

The thermopower as well as the conductivity give a measure of the mass at the Fermi level, while the Zeeman field B_c required to fully polarize the spins measures the mass related to the bandwidth, which is the Fermi energy counted from the band bottom. For $n_s \geq 10^{11}$ cm^{-2}, the mass determined by different methods was found to be essentially the same (Kravchenko and Sarachik, 2004; Shashkin, 2005). On the other hand, the behavior is different at the densities reached in the experiment in the very close vicinity of the critical point n_t ($n_s < 10^{11}$ cm^{-2}) where the bandwidth-related mass was found to increase by only a factor of $\sim$4. This follows from the fact that the Shubnikov–de Haas oscillations in the dilute 2D electron system in silicon reveal one switch from cyclotron to spin minima (the ratio of the spin and cyclotron splittings reaches $\sim$1) as the electron density is decreased (Kravchenko et al., 2000), the spin minima surviving down to $n_s \approx n_c$ and even below (D'Iorio et al., 1990). In effect, while the bandwidth does not decrease appreciably in the close vicinity of the critical point n_t and the effective mass obtained from such measurements does not exhibit a true divergence, the thermopower measurements yield the effective mass at the Fermi energy, which does indeed diverge.

A divergence of the effective mass has been predicted by a number of theories: using Gutzwiller's theory (Dolgopolov, 2002); using an analogy with He^3 near the onset of Wigner crystallization (Spivak, 2003; Spivak and Kivelson, 2004); extending the Fermi liquid concept to the strongly interacting limit (Khodel et al., 2008); solving an extended Hubbard model using dynamical mean-field theory (Pankov and Dobrosavljevic, 2008); from a renormalization group analysis for multivalley 2D systems (Punnoose and Finkelstein, 2005); by Monte Carlo simulations (Fleury and Waintal, 2010; Marchi et al., 2009). Some theories predict that the disorder is important for the mass enhancement (Fleury and Waintal, 2010; Marchi et al., 2009; Punnoose and Finkelstein, 2005). In contrast with most theories that assume a parabolic spectrum, the authors

Khodel et al. (2008) stress that there is a clear distinction between the mass at the Fermi level and the bandwidth-related mass. In this respect, our conclusions are consistent with the model of Khodel et al. (2008) in which a flattening at the Fermi energy in the spectrum leads to a diverging effective mass. This Fermi liquid–based model implies the existence of an intermediate phase that precedes Wigner crystallization.

There has been a great deal of debate concerning the origin of the interesting, enigmatic behavior in these strongly interacting 2D electron systems. In particular, many have questioned whether the change of the resistivity from metallic to insulating temperature dependence signals a phase transition, or whether it is a crossover. We close by noting that, unlike the resistivity which displays complex behavior that may not distinguish between these two scenarios, we have shown by measurements of the spin susceptibility, effective mass, and thermopower that the 2D electron system in low-disordered silicon MOSFETs behaves critically at a well-defined density, providing clear evidence that this is a transition to a new phase at low densities. The next challenge is to determine the nature of this phase.

References

Abrahams, E., Anderson, P. W., Licciardello, D. C., and Ramakrishnan, T. V. (1979). *Phys. Rev. Lett.* **42**, pp. 673–676.

Abrahams, E., Kravchenko, S. V., and Sarachik, M. P. (2001). *Rev. Mod. Phys.* **73**, pp. 251–266.

Adkins, C. J., Pollitt, S., and Pepper, M. (1976). *J. Phys. C* **37**, pp. 343–347.

Altshuler, B. L., Aronov, A. G., and Lee, P. A. (1980). *Phys. Rev. Lett.* **44**, pp. 1288–1291.

Anissimova, S., Venkatesan, A., Shashkin, A. A., Sakr, M. R., Kravchenko, S. V., and Klapwijk, T. M. (2006). *Phys. Rev. Lett.* **96**, p. 046409.

Attaccalite, C., Moroni, S., Gori-Giorgi, P., and Bachelet, G. B. (2002). *Phys. Rev. Lett.* **88**, p. 256601.

Castellani, C., Di Castro, C., Lee, P. A., and Ma, M. (1984). *Phys. Rev. B* **30**, pp. 527–543.

D'Iorio, M., Pudalov, V. M., and Semenchinsky, S. G. (1990). *Phys. Lett. A* **150**, pp. 422–426.

Dolgopolov, V. T., Kravchenko, G. V., Shashkin, A. A., and Kravchenko, S. V. (1992). *Phys. Rev. B* **46**, pp. 13303–13308.

Dolgopolov, V. T. and Gold, A. (2000). *JETP Lett.* **71**, pp. 27–30.

Dolgopolov, V. T. (2002). *JETP Lett.* **76**, pp. 377–379.

Dolgopolov, V. T. and Gold, A. (2011). *JETP Lett.* **94**, pp. 446–450.

Faniel, S., Moldovan, L., Vlad, A., Tutuc, E., Bishop, N., Melinte, S., Shayegan, M., and Bayot, V. (2007). *Phys. Rev. B* **76**, p. 161307(R).

Finkelstein, A. M. (1983). *Sov. Phys. JETP* **57**, pp. 97–108.

Finkelstein, A. M. (1984). *Z. Phys. B* **56**, pp. 189–196.

Fletcher, R., Pudalov, V. M., Feng, Y., Tsaousidou, M., and Butcher, P. N. (1997). *Phys. Rev. B* **56**, pp. 12422–12428.

Fleury, G. and Waintal, X. (2010). *Phys. Rev. B* **81**, p. 165117.

Gold, A. and Dolgopolov, V. T. (2011). *Europhys. Lett.* **96**, p. 27007.

Goswami, S., Siegert, C., Baenninger, M., Pepper, M., Farrer, I., Ritchie, D. A., and Ghosh, A. (2009). *Phys. Rev. Lett.* **103**, p. 026602.

Jaroszyński, J., Popović, D., and Klapwijk, T. M. (2002). *Phys. Rev. Lett.* **89**, p. 276401.

Khodel, V. A., Clark, J. W., and Zverev, M. V. (2008). *Phys. Rev. B* **78**, p. 075120.

Khrapai, V. S., Shashkin, A. A., and Dolgopolov, V. T. (2003). *Phys. Rev. Lett.* **91**, p. 126404.

Kravchenko, S. V., Kravchenko, G. V., Furneaux, J. E., Pudalov, V. M., and D'Iorio, M. (1994). *Phys. Rev. B* **50**, pp. 8039–8042.

Kravchenko, S. V., Mason, W. E., Bowker, G. E., Furneaux, J. E., Pudalov, V. M., and D'Iorio, M. (1995). *Phys. Rev. B* **51**, pp. 7038–7045.

Kravchenko, S. V., Shashkin, A. A., Bloore, D. A., and Klapwijk, T. M. (2000). *Solid State Commun.* **116**, pp. 495–499.

Kravchenko, S. V. and Klapwijk, T. M. (2000). *Phys. Rev. Lett.* **84**, pp. 2909–2912.

Kravchenko, S. V. and Sarachik, M. P. (2004). *Rep. Prog. Phys.* **67**, pp. 1–44.

Landau, L. D. (1957). *Sov. Phys. JETP* **3**, pp. 920–925.

Marchi, M., De Palo, S., Moroni, S., and Senatore, G. (2009). *Phys. Rev. B* **80**, p. 035103.

Mason, W., Kravchenko, S. V., Bowker, G. E., and Furneaux, J. E. (1995). *Phys. Rev. B* **52**, pp. 7857–7859.

Mokashi, A., Li, S., Wen, B., Kravchenko, S. V., Shashkin, A. A., Dolgopolov, V. T., and Sarachik, M. P. (2012). *Phys. Rev. Lett.* **109**, p. 096405.

Okamoto, T., Hosoya, K., Kawaji, S., and Yagi, A. (1999). *Phys. Rev. Lett.* **82**, pp. 3875–3878.

Pankov, S. and Dobrosavljević, V. (2008). *Phys. Rev. B* **77**, p. 085104.

Polyakov, D. G. and Shklovskii, B. I. (1993). *Phys. Rev. B* **48**, pp. 11167–11175.

Pudalov, V. M., D'Iorio, M., Kravchenko, S. V., and Campbell, J. W. (1993). *Phys. Rev. Lett.* **70**, pp. 1866–1869.

Punnoose, A. and Finkelstein, A. M. (2005). *Science* **310**, pp. 289–291.

Sarachik, M. P. and Kravchenko, S. V. (1999). *Proc. Natl. Acad. Sci. USA* **96**, pp. 5900–5902.

Shashkin, A. A., Dolgopolov, V. T., and Kravchenko, G. V. (1994a). *Phys. Rev. B* **49**, pp. 14486–14495.

Shashkin, A. A., Dolgopolov, V. T., Kravchenko, G. V., Wendel, M., Schuster, R., Kotthaus, J. P., Haug, R. J., von Klitzing, K., Ploog, K., Nickel, H., and Schlapp, W. (1994b). *Phys. Rev. Lett.* **73**, pp. 3141–3144.

Shashkin, A. A., Kravchenko, S. V., and Klapwijk, T. M. (2001a). *Phys. Rev. Lett.* **87**, p. 266402.

Shashkin, A. A., Kravchenko, S. V., Dolgopolov, V. T., and Klapwijk, T. M. (2001b). *Phys. Rev. Lett.* **87**, p. 086801.

Shashkin, A. A., Kravchenko, S. V., Dolgopolov, V. T., and Klapwijk, T. M. (2002). *Phys. Rev. B* **66**, p. 073303.

Shashkin, A. A., Rahimi, M., Anissimova, S., Kravchenko, S. V., Dolgopolov, V. T., and Klapwijk, T. M. (2003). *Phys. Rev. Lett.* **91**, p. 046403.

Shashkin, A. A. (2005). *Phys. Usp.* **48**, pp. 129–149.

Shashkin, A. A., Anissimova, S., Sakr, M. R., Kravchenko, S. V., Dolgopolov, V. T., and Klapwijk, T. M. (2006). *Phys. Rev. Lett.* **96**, p. 036403.

Shashkin, A. A., Kapustin, A. A., Deviatov, E. V., Dolgopolov, V. T., and Kvon, Z. D. (2007). *Phys. Rev. B* **76**, p. 241302(R).

Spivak, B. (2003). *Phys. Rev. B* **67**, p. 125205.

Spivak, B. and Kivelson, S. A. (2004). *Phys. Rev. B* **70**, p. 155114.

Stoner, E. C. (1946). *Rep. Prog. Phys.* **11**, pp. 43–112.

Tanatar, B. and Ceperley, D. M. (1989). *Phys. Rev. B* **39**, pp. 5005–5016.

Wigner, E. (1934). *Phys. Rev.* **46**, pp. 1002–1011.

Vitkalov, S. A., Zheng, H., Mertes, K. M., Sarachik, M. P., and Klapwijk, T. M. (2000). *Phys. Rev. Lett.* **85**, pp. 2164–2167.

Zala, G., Narozhny, B. N., and Aleiner, I. L. (2001). *Phys. Rev. B* **64**, p. 214204.

Zavaritskaya, T. N. and Zavaritskaya, E. I. (1987). *JETP Lett.* **45**, pp. 609–613.

Chapter 3

Transport in a Two-Dimensional Disordered Electron Liquid with Isospin Degrees of Freedom

Igor S. Burmistrov

L.D. Landau Institute for Theoretical Physics RAS, Kosygina str. 2,
Moscow 119334, Russia
burmi@itp.ac.ru

Recent theoretical results on transport in a two-dimensional disordered electron liquid with spin and isospin degrees of freedom are reviewed. A number of experimental features in temperature dependence of resistivity at low temperatures in a Si-MOSFET, an n-AlAs quantum well, and double-quantum-well heterostructures is explained. Novel behaviors of low-temperature resistivity at low temperature in electron systems with isospin degrees of freedom are predicted.

3.1 Introduction

The essence of the phenomenon of Anderson localization is quantum interference that can fully suppress diffusion of a quantum

Strongly Correlated Electrons in Two Dimensions
Edited by Sergey Kravchenko
Copyright © 2017 Pan Stanford Publishing Pte. Ltd.
ISBN 978-981-4745-37-6 (Hardcover), 978-981-4745-38-3 (eBook)
www.panstanford.com

particle in a random potential (Anderson, 1958). The wave function of a particle with a given energy can be localized or delocalized in space. It leads to a possibility for existence of quantum phase transition with changing particle energy or parameters of the random potential. This quantum phase transition is usually termed as Anderson transition. Instead of quantum mechanical problem for a single particle one can consider a system of noninteracting electrons. Then the Anderson transition can occur under change of the chemical potential or dimensionless parameter $k_F l$ where k_F denotes the Fermi momentum and l stands for the elastic mean free path. From the point of view of transport properties the phase with delocalized states is a metal whereas the phase with localized states is an insulator.

In general, three scenarios are possible: (i) the states at all energies are localized, (ii) the states at all energies are delocalized, and (iii) the mobility edge between localized and delocalized states (Anderson transition) exists. In $d = 1$ dimension all states are localized in any random potential (Mott and Twose, 1961); see also Fröhlich et al. (1985). In $d = 3$ mobility edge and Anderson transition exists (Anderson, 1958). Existence of Anderson transition in $d = 2$ is more complicated question. Based on the relation of conductance to the response to change in the boundary conditions for a finite-size system (Thouless, 1974), the scaling theory for conductance has been constructed (Abrahams et al., 1979). This scaling theory was in agreement with direct diagrammatic calculations of conductance in weak disorder limit ($k_F l \gg 1$) (Gorkov et al., 1979; Abrahams and Ramakrishnan, 1980). The scaling theory of Anderson and coworkers (Abrahams et al., 1979) predicts that in $d = 2$ all electronic states are localized. In $d > 2$ Anderson transition occurs and conductance vanishes as a power law at the transition point (Wegner, 1976). The scaling theory of Anderson transition allows to use for its description powerful tools developed for critical phenomena: low-energy effective action and renormalization group (RG) (see, for example, Amit, 1984; Zinn-Justin, 1989).

For the problem of Anderson localization low-energy effective action is the so-called nonlinear σ-model (NLσM) (Wegner, 1979; Schäfer and Wegner, 1980; Efetov et al., 1980; Jüngling and Oppermann, 1980; McKane and Stone, 1981; Efetov, 1982). It describes diffusive motion of electrons on scales larger than the mean free path.[a] The study of Anderson localization in the limit of weak disorder is much more convenient with the help of NLσM than using standard diagrammatic technique (Lee and Ramakrishnan, 1985; Efetov, 1983). The former describes the interaction of diffusive modes, leading to length scales larger than the elastic mean free path to logarithmic divergences in $d = 2$.[b]

The existence of Anderson transition depends not only on spatial dimensionality but also on symmetry of single-particle Hamiltonian. In $d = 2$ Anderson transition is absent for orthogonal Wigner–Dyson symmetry class (Wigner, 1951; Dyson, 1962a, 1962b), that is, for Hamiltonians that preserve time reversal and spin-rotational symmetries. For example, in the case of symplectic symmetry class in which spin-rotational symmetry is broken Anderson transition occurs in $d = 2$. In general, there are 10 different symmetry classes of single-particle Hamiltonians describing (quasi)particle motion in a random potential (Zirnbauer, 1996; Altland and Zirnbauer, 1997; Heinzner et al., 2005).

For each symmetry class the corresponding NLσM contains the topological terms in some spatial dimensions (Schnyder et al., 2008; Schnyder et al., 2009; Kitaev, 2009). Their existence affects localization crucially. The textbook example is integer quantum Hall effect. The Hamiltonian for a particle in $d = 2$ in the presence of perpendicular magnetic field which breaks time reversal symmetry corresponds to the unitary class. Standard scaling theory in the unitary class predicts localization at large length scales (Brézin et al., 1980). Existence of the topological term (Levine et al., 1983; Pruisken, 1984) in NLσM for the unitary class in $d = 2$ results in delocalized states and integer quantized Hall conductance (Pruisken, 1987).

---

[a] In $d = 1$ for the single-channel case the NLσM approach is not applicable since localization length is given by the mean free path and there is no spatial scales at which diffusive motion exists (Berezinskii, 1974).

[b] Diffusive modes include diffusons (two-particle propagator in a particle–hole channel) and cooperons (two-particle propagator in a particle–particle channel).

We emphasize that different aspects of Anderson localization discussed above exist in the problem of noninteracting electrons.[c] To realize a system of noninteracting electrons in a laboratory is not an easy task since one needs to exclude electron–electron and electron–phonon interactions.[d] At low temperatures electron–electron interaction becomes important. Inelastic processes due to electron–electron scattering with small (compared to temperature) energy transfer lead to destruction of phase coherence on timescales longer than the phase-breaking time τ_ϕ (Thouless, 1977; Abrahams et al., 1979; Altshuler et al., 1982).[e] As temperature is lowered the phase-breaking time τ_ϕ and corresponding spatial scale L_ϕ increase and diverge at zero temperature. At finite temperatures such that $L_\phi < L$, the phase-breaking length plays a role of an effective system size. It yields the temperature dependence of conductance (Altshuler et al., 1982).

In addition to influence on conductance through the phase-breaking time,[f] electron–electron interaction results in strong temperature dependence of conductance at low temperatures due to virtual processes in electron–electron scattering (Altshuler and Aronov, 1979a). Physically, strong temperature dependence of conductance is due to coherent scattering electrons off the Fridel oscillations (Zala et al., 2001a). Strong (as compared with Fermi liquid) temperature dependence appears also in thermodynamic quantities, for example, the specific heat and the static spin susceptibility (see, for example, Altshuler and Aronov, 1985). The most interesting case is the case of $d = 2$ in which both weak localization and electron–electron (Altshuler et al., 1980)

---

[c]We refer the reader to Mirlin (2000), Mirlin and Evers (2008), and Dobrosavljević (2010) for a detailed exposition of the current progress in the studies of Anderson localization.

[d]Recently, experiments on Anderson localization in cold atoms in random optical potential have been performed (see, for example, Jendrzejewski et al., 2012; Lopez et al., 2012).

[e]We note that in general phase-breaking rate is different from the rate of electron–electron collisions (out-scattering rate in the kinetic equation approach [Schmidt, 1974; Altshuler and Aronov, 1979b]) (see, for example, Blanter, 1996).

[f]Provided in the noninteracting approximation all single-particle states are localized, electron–electron interaction can lead to the finite-temperature transition between the phase with zero (low-temperature phase) and finite (high-temperature phase) conductivity (Fleishman and Anderson, 1980; Gornyi et al., 2005; Basko et al., 2006).

contributions to conductance are logarithmic in temperature. In the orthogonal symmetry class, contribution to the conductance due to strong electron–electron interaction is of opposite sign with respect to weak localization contribution. Therefore, in the presence of electron–electron interaction the metal–insulator quantum phase transition is possible in $d = 2$.

The first attempt to extend the scaling theory of Anderson and coworkers (Abrahams et al., 1979) to the metal–insulator transition in the presence of electron–electron interaction was performed by McMillan (McMillan, 1981). In spite of confusion between the local density of states and the thermodynamic one, important outcome of McMillan (McMillan, 1981) was an idea of two-parameter scaling description for metal–insulator transition in the presence of interaction. The breakthrough was done by Finkelstein (Finkelstein, 1983a), where NLσM was derived for the case of interacting electrons starting from the microscopic theory. With the help of RG analysis of this, so-called Finkelstein NLσM, the scaling theory of the metal–insulator transition in the presence of electron–electron interaction for $d > 2$ was built (Finkelstein, 1983b; Finkelstein, 1984a, 1984b, 1984c; Castellani et al., 1984a, 1984b). As usual, strong electron–electron interaction results in a change of the universality class of the metal–insulator transition in comparison with the noninteracting case (see, for example, Finkelstein, 1990; Belitz and Kirkpatrick, 1994). In other words, electron–electron interaction is usually a relevant perturbation. We mention that recently the influence of electron–electron interaction on localization has been studied for the superconducting and chiral symmetry classes (Dell'Anna, 2006).

In $d = 2$ for the orthogonal symmetry class in the range of weak disorder the contribution to the conductance due to electron–electron interaction is stronger than weak localization contribution. In total, this leads to metallic behavior of conductivity at low temperatures (Finkelstein, 1984c). This fact is in favor of existence of metal–insulator transition in $d = 2$ in the presence of electron–electron interaction.

In spite of long history for experimental research of 2D electron systems (Ando et al., 1982), experimental observation of change in resistivity from insulating to metallic behavior with increase of

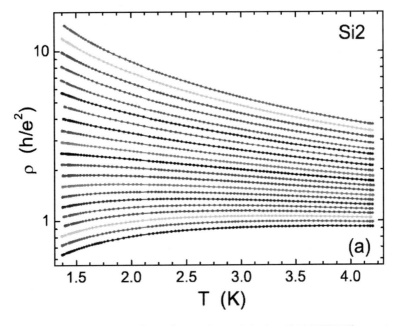

Figure 3.1 Temperature dependence of resistivity in a Si-MOSFET. Electron concentration increases from top to bottom on 0.224×10^{10} cm^{-2}, starting from 6.72×10^{10} cm^{-2} for the upper curve. Reprinted (figure) with permission from [Knyazev, D. A., Omel'yanovskii, O. E., Pudalov, V. M., and Burmistrov, I. S. (2008). *Phys. Rev. Lett.* **100**, p. 46405.] Copyright (2008) by the American Physical Society.

electron density in a Si metal-oxide-semiconductor field-effect transistor (Si-MOSFET) (see Fig. 3.1) became a surprise (Kravchenko et al., 1994; Kravchenko et al., 1995). The observed temperature and electron density dependence of resistivity is as expected for resistivity in the vicinity of the metal–insulator transition. Later similar behavior of resistivity was experimentally observed in a variety of 2D electron systems (Abrahams et al., 2001; Shangina and Dolgopolov, 2003; Kravchenko and Sarachik, 2004; Pudalov et al., 2004; Shashkin, 2005; Gantmakher and Dolgopolov, 2008). For the observed (Kravchenko et al., 1994; Kravchenko et al., 1995) temperature behavior of resistivity an important role seems to be played by the electron spin. An indication for this conclusion is the following experimental fact for a Si-MOSFET (see Fig. 3.2): a

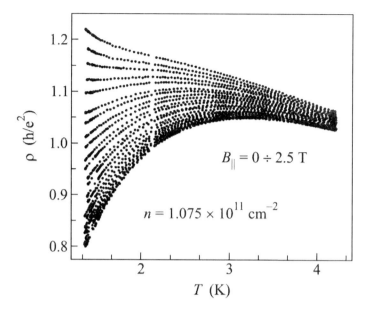

Figure 3.2 Temperature dependence of resistivity in a Si-MOSFET with electron concentration $n = 1.075 \times 10^{11}$ cm^{-2} in a parallel magnetic field. The magnetic field is changed on 0.1 T from 0 (lower curve) to 2.5 (upper curve). From Knyazev, D. A., Omel'yanovskii, O. E., Pudalov, V. M., and Burmistrov, I. S. (2006). *JETP Lett.* **84**, p. 662. With permission of Springer.

weak (as compared with the field necessary for the full polarization) parallel magnetic field changes behavior of the resistivity from metallic to insulating (Simonian et al., 1997; Vitkalov et al., 2003; Pudalov et al., 2003).

Similar effect of a parallel magnetic field was observed experimentally in a 2D hole system in GaAs/AlGaAs heterostructure (Yoon et al., 2000). In the latter case, such behavior of resistivity with a parallel magnetic field can be explained by scaling theory, which takes into account that due to Zeeman splitting two among three triplet diffusive modes become massive (Finkelstein, 1990). Recently, the existence of the metal–insulator transition was demonstrated theoretically in $d = 2$ for the special case in which the electron spin is equal to $\mathcal{N} \to \infty$ (the action of NLσM is invariant under $SU(\mathcal{N})$ rotations) (Punnoose and Finkelstein, 2005). It should be contrasted with the case of spinless electrons with Coulomb interaction for which the metal–insulator transition is absent in $d = 2$ (Baranov, 2002). Therefore, spin degrees of freedom can crucially affect the low-temperature behavior of resistivity and existence of the metal–insulator transition.

Recently, a detailed comparison of the two-parameter scaling theory of Finkelstein (Finkelstein, 1990) with experimental data in a Si-MOSFET has been performed. Temperature dependence of resistivity in the metallic region (Punnoose and Finkelstein, 2001), magnetoresistance in a parallel field (Knyazev et al., 2007; Anissimova et al., 2007), and resistance at criticality (Knyazev et al., 2008) demonstrate reasonable agreement with the theory. We emphasize that 2D electron system in a Si-MOSFET is specific since electrons occupy two valleys.[g] This leads to appearance of two additional energy scales: valley splitting Δ_v and intervalley elastic scattering rate $1/\tau_v$. In the vicinity of the critical region they were estimated as $\Delta_v \approx 1$ K and $1/\tau_v \approx 0.1$ K (Kuntsevich et al., 2007; Klimov et al., 2008). As a consequence the theory which takes into account finite values of Δ_v and $1/\tau_v$ should be developed for detailed comparison with experimental data on transport below 1 K in a two-valley electron system such as a Si-MOSFET.

We mention that recently 2D two-valley electron system in $SiO_2/Si(100)/SiO_2$ quantum well has been experimentally studied (Takashina et al., 2011; Renard et al., 2013). However, Renard et al. (2013) estimated the intervalley scattering rate to be of the order of 7 K such that the 2D electron system behaves as the single-valley one for transport at low temperatures ($T \lesssim 7$ K).

It is well known (Shayegan et al., 2006) that a two-valley electron system can be also realized in an n-AlAs quantum well.[h] Contrary

---

[g] In 3D Si the Brillouin zone contains six valleys which are degenerate in energy. In a Si(001)-MOSFET degeneracy is lifted and only two valleys with centers at the z axis in the reciprocal space remain at low energies. The other four valleys are separated by a gap of the order of 200 K and, therefore, do not participate in low-temperature transport.

[h] In 3D AlAs there are six valley which are degenerate in energy. Their centers are exactly at the boundaries of the Brillouin zone such that there are three valleys per each zone. In an n-AlAs quantum well degeneracy is partially lifted, and only the valleys with centers at x and y axes in the reciprocal space contribute to low-temperature transport. Therefore, there are effectively two valleys per Brllouin zone that participate in low-temperature transport in an n-AlAs quantum well.

to a Si-MOSFET, in an n-AlAs quantum well one can affect not only spin degrees of freedom by applying a parallel magnetic field but also isospin degrees of freedom (due to two valleys) also by means of stress that controls valley splitting (Gunawan et al., 2006, 2007). Gunawan et al. (2006, 2007). found that to change temperature dependence of resistivity from metallic to insulating it is not enough to apply either a parallel magnetic field or stress only. We mention that this experimental fact is in agreement with the experiments in a Si-MOSFET (Simonian et al., 1997; Vitkalov et al., 2003; Pudalov et al., 2003). Indeed in a Si-MOSFET finite valley splitting is built in. However, these experimental results contradict the theory of Burmistrov and Chtchelkatchev (2007) with straightforward generalization of the theory of Punnoose and Finkelstein (2001) to the case of finite spin or valley splitting only. According to results of Burmistrov and Chtchelkatchev (2007) spin or valley splitting only is enough to change behavior of the resistivity from metallic to insulating. The theoretical results reviewed in this chapter resolve this discrepancy between theory and experiment.

One more system of 2D electrons which has isospin degrees of freedom is a double-quantum-well heterostructure. In this case isospin distinguishes states localized in different quantum wells. In spite of a large number of interesting phenomena in double-quantum-well heterostructures, such as Coulomb drag (Gramila et al., 1991; Sivan et al., 1992; Lilly et al., 1998; Pillarisetty et al., 2002), exciton condensation (Kellog et al., 2004; Tutuc et al., 2004; Eisenstein and MacDonald, 2004), ferromagnetic (Boebinger et al., 1990; Sawada et al., 1999) and counted antiferromagnetic (Khrapai et al., 2000) phases, low-temperature electron transport in the vicinity of possible metal–insulator transition has not been studied experimentally in detail. Electron transport was studied only at large electron concentrations in the metallic region (Minkov et al., 2000; Pagnossin et al., 2008).

In a double-quantum-well heterostructure with common scatterers for electrons one can expect behavior of resistivity similar to the one in a Si-MOSFET. Control of electron concentration in one quantum well by gate voltage allows to interpolate between the case of equal electron concentrations in each quantum well (akin to a two-valley system) and the case in which one quantum well is empty (akin to the single-valley system). According to the theory of Punnoose and Finkelstein (2001), one can expect strong change in temperature dependence of resistivity in the case of transfer from a single- to a two-valley situation. However, recent experiments (Minkov et al., 2010, 2011) in double-quantum-well heterostructures do not demonstrate a significant difference in temperature dependence of resistivity during exhaustion of one of quantum wells. The theory reviewed in this chapter allows us to explain this unexpected behavior of the resistivity measured by Minkov et al. (2010, 2011).

The chapter is organized as follows. In Section 3.2 we remind readers of the NLσM approach and derive general RG equations in one-loop approximation in the case when an electron is characterized by both spin and isospin. In Section 3.3 we use our general results for the description of low-temperature transport in a 2D two-valley electron system in a Si-MOSFET in the presence of spin and valley splittings. In Section 3.4, using our general results we explain transport experiments in a 2D electron system in a double-quantum-well heterostructure with common scatterers. We end the chapter with conclusions (Section 3.5).

3.2 Nonlinear σ-Model

3.2.1 Nonlinear σ-Model Action

At low temperatures $T\tau_{tr} \ll 1$, where τ_{tr} denotes transport elastic scattering time, NLσM is the convenient tool for description of disordered electron liquid. NLσM is designed to follow the interaction of low-energy ($|E| \lesssim 1/\tau_{tr}$) bosonic modes termed as diffusons and cooperons. NLσM is the field theory of the matrix field $Q_{mn}^{\alpha\beta}(\mathbf{r})$ which satisfies the following constraints: $Q^2(\mathbf{r}) = 1$, $\mathrm{tr}\,Q(\mathbf{r}) = 0$, $Q^+(\mathbf{r}) = Q(\mathbf{r})$. The Greek indices α, $\beta = 1, 2, \ldots, N_r$ stand for the replica indices, whereas the Latin indices are integers m, n corresponding to Matsubara energies $\varepsilon_n = \pi T(2n + 1)$. In general, the matrix element $Q_{mn}^{\alpha\beta}(\mathbf{r})$ can have a matrix structure and satisfies additional constraints. The symbol tr denotes the trace-over replica, Matsubara, spin, and isospin indices. We refer the reader to

Finkelstein and to Belitz and Kirkpatrick (Finkelstein, 1990; Belitz and Kirkpatrick, 1994) for a review of the NLσM approach. In what follows we assume the presence of a weak perpendicular magnetic field $B \gtrsim \max\{1/\tau_\phi, T\}/D$ where D denotes the diffusion coefficient (see, for example, Altshuler and Aronov, 1985) such that cooperons and interaction in the Cooper channel are suppressed. In this case $Q_{mn}^{\alpha\beta}(\mathbf{r})$ is a 4×4 matrix in spin and isospin spaces.

The action of NLσM is as follows:

$$S = S_\sigma + S_F + S_{SB}. \tag{3.1}$$

Here the first term of the right-hand side (S_σ) describes NLσM for the noninteracting electrons (Wegner, 1979; Schäfer and Wegner, 1980; Efetov et al., 1980; McKane and Stone, 1981):

$$S_\sigma = -\frac{\sigma_{xx}}{32} \int d\mathbf{r} \, \mathrm{tr}(\nabla Q)^2, \tag{3.2}$$

where $\sigma_{xx} = 4\pi \nu_\star D$ stands for the Drude conductivity in units e^2/h. The thermodynamic density of states $\nu_\star = m_\star/\pi$ is determined by the effective mass $m_\star$ (it includes Fermi liquid corrections). We mention that the dimensionless Drude conductivity is given by $\sigma_{xx} = 2k_F l$, and it is assumed that $\sigma_{xx} \gg 1$.

The presence of electron–electron interaction yields additional contribution to the NLσM action (Finkelstein, 1983a, 1983b, 1984a, 1984b; Pruisken et al., 1999):

$$S_F = -\pi T \int d\mathbf{r} \left[\sum_{\alpha n; ab} \frac{\Gamma_{ab}}{4} \, \mathrm{tr} \, I_n^\alpha t_{ab} Q(\mathbf{r}) \, \mathrm{tr} \, I_{-n}^\alpha t_{ab} Q(\mathbf{r}) \right.$$

$$\left. -4z \, \mathrm{tr} \, \eta Q + 6z \, \mathrm{tr} \, \eta \Lambda \right]. \tag{3.3}$$

Here 16 matrices $t_{ab} = \tau_a \otimes \sigma_b$ ($a, b = 0, 1, 2, 3$) are the generators of the $SU(4)$ group. The Pauli matrices τ_a, $a = 0, 1, 2, 3$, acts on the isospin indices but the Pauli matrices σ_b, $b = 0, 1, 2, 3$, operates in the spin space. The quantities Γ_{ab} stand for the interaction amplitudes. The parameter z is independent of the charge (in the field theory sense) of NLσM. It was introduced originally by Finkelstein (1983a) in order for the RG flow to be consistent with the particle number conservation. Its bare value is determined by the thermodynamic density of states: $z = \pi\nu_\star/4$. Physically, z governs the temperature dependence of the specific heat (Castellani and Di Castro, 1986). The interaction amplitudes Γ_{ab} in the action Eq. 3.1 are related with the Fermi liquid interaction parameters F_{ab}: $\Gamma_{ab} = -zF_{ab}/(1 + F_{ab})$. Matrices Λ, η and I_k^γ are defined as follows:

$$\Lambda_{nm}^{\alpha\beta} = \operatorname{sign} n\delta_{nm}\delta^{\alpha\beta} t_{00}, \qquad \eta_{nm}^{\alpha\beta} = n\delta_{nm}\delta^{\alpha\beta} t_{00},$$
$$(I_k^\gamma)_{nm}^{\alpha\beta} = \delta_{n-m,k}\delta^{\alpha\gamma}\delta^{\beta\gamma} t_{00}. \qquad (3.4)$$

The last term in the right-hand side of Eq. 3.1 describes the effect of symmetry-breaking terms:

$$\mathcal{S}_{SB} = iz_{ab}\Delta_{ab} \int d\mathbf{r} \operatorname{tr} t_{ab}Q + \frac{N_r z_{ab}}{\pi T} \int d\mathbf{r} \, \Delta_{ab}^2. \qquad (3.5)$$

For example, the Zeeman splitting Δ_s yields the term Eq. 3.5 with $t_{ab} = t_{03}$ and $\Delta_{03} = \Delta_s$. In the case of the two-valley 2D electron system in a Si-MOSFET, the valley splitting Δ_v results in $\Delta_{30} = \Delta_v$ in Eq. 3.5.

Matrix Q in action Eq. 3.1 is formally of the infinite size in the Matsubara space. However, to handle it one needs to introduce cutoff at large Matsubara frequencies. We assume that integers m, n which correspond to Matsubara energies are restricted to the range $-N_M \leqslant m, n \leqslant N_M - 1$ with $N_M \gg 1$. The condition of applicability of NLσM gives the following estimate $N_M \approx 1/(2\pi T \tau_{tr})$.

It is well known (Pruisken et al., 1999; Kamenev and Andreev, 1999) that global rotations of Q by matrix $\exp(i\hat{\chi})$:

$$Q(\mathbf{r}) \to \exp(i\hat{\chi})Q(\mathbf{r})\exp(-i\hat{\chi}), \quad \hat{\chi} = \sum_{\alpha,n} \chi_n^\alpha I_n^\alpha \qquad (3.6)$$

are important due to their relation with spatially constant electric potential. The latter can be gauged away by suitable gauge transformation of electron operators. Therefore it is convenient to define the limit $N_M \to \infty$ in a such way that the following relations hold ($\mathcal{F}$-algebra) (Pruisken et al., 1999):[i]

---

[i]We mention that the limit $N_M \to \infty$ should be taken at fixed $N = N_r N_M$. Then, the latter vanishes in the replica limit $N_r \to 0$. NLσM with a finite value of N_M is equivalent to NLσM for noninteracting electrons at a length scale exceeding $\sqrt{D/2\pi T N_M}$ (Pruisken et al., 2005).

$$\text{tr } I_n^\alpha t_{ab} e^{i\hat{\chi}} Q e^{-i\hat{\chi}} = \text{tr } I_n^\alpha t_{ab} e^{i\chi_0} Q e^{-i\chi_0} + 8in(\chi_{ab})_{-n}^\alpha ,$$

$$\text{tr } \eta e^{i\hat{\chi}} Q e^{-i\hat{\chi}} = \text{tr } \eta Q + \sum_{\alpha n; ab} in(\chi_{ab})_n^\alpha \text{ tr } I_n^\alpha t_{ab} Q$$

$$-4 \sum_{\alpha n; ab} n^2 (\chi_{ab})_n^\alpha (\chi_{ab})_{-n}^\alpha. \tag{3.7}$$

Here $\chi_0 = \sum_\alpha \chi_0^\alpha I_0^\alpha$. With the help of the relations in Eq. 3.7 one can check that provided $\Gamma_{00} = -z$ the NLσM action $S_\sigma + S_F$ is invariant under global rotations, Eq. 3.6, with $\chi_{ab} = \chi \delta_{a0} \delta_{b0}$.

3.2.2 Physical Observables

The most important quantities that contain information on low-energy behavior of NLσM are physical observables σ'_{xx}, z', and z'_{ab}. They correspond to the parameters σ_{xx}, z, and z_{ab} of action Eq. 3.1. Here σ'_{xx} is the conductivity of electron system defined via linear response to an external electromagnetic field. The observable z' determines the specific heat (Castellani and Di Castro, 1986) and z'_{ab} corresponds to the static generalized susceptibility (Finkelstein, 1984b; Castellani et al., 1984), $\chi_{ab} = 2z'_{ab}/\pi$, which are responses to Δ_{ab}.

The conductivity σ'_{xx} is given by the Kubo formula (Baranov et al., 1999)

$$\sigma'_{xx}(i\omega_n) = -\frac{\sigma_{xx}}{16n} \left\langle \text{tr}[I_n^\alpha, Q][I_{-n}^\alpha, Q] \right\rangle + \frac{\sigma_{xx}^2}{64dn} \int d\mathbf{r}'$$

$$\times \langle\langle \text{tr } I_n^\alpha Q(\mathbf{r}) \nabla Q(\mathbf{r}) \times \text{tr } I_{-n}^\alpha Q(\mathbf{r}') \nabla Q(\mathbf{r}') \rangle\rangle. \tag{3.8}$$

Here the analytic continuation to the real frequencies, $i\omega_n \to \omega + i0^+$ and static limit $\omega \to 0$ are should be taken. The averages in Eq. 3.8 are with respect to the action Eq. 3.1, and $\langle\langle A \times B \rangle\rangle = \langle AB \rangle - \langle A \rangle \langle B \rangle$. The physical observable z' is determined by the thermodynamic potential Ω (Baranov et al., 1999):

$$z' = \frac{1}{2\pi \text{ tr } \eta \Lambda} \frac{\partial}{\partial T} \frac{\Omega}{T}. \tag{3.9}$$

The physical observables z'_{ab} can be found from the following relations (Finkelstein, 1990):

$$z'_{ab} = \frac{\pi}{2N_r} \frac{\partial^2 \Omega}{\partial \Delta_{ab}^2}. \tag{3.10}$$

3.2.3 One-Loop Renormalization

3.2.3.1 Perturbation theory

To construct perturbation theory in a small parameter $1/\sigma_{xx}$ it is convenient to use the square-root parameterization of Q:

$$Q = W + \Lambda \sqrt{1 - W^2}, \qquad W = \begin{pmatrix} 0 & w \\ w & 0 \end{pmatrix}. \tag{3.11}$$

Then Eq. 3.1 can be written as an infinite series in powers of w and w fields. In the absence of the symmetry-breaking term $\mathcal{S}_{SB}$ the propagator of fields w and $\bar{w}$ becomes:

$$\langle [w_{ab}(\mathbf{q})]_{n_1 n_2}^{\alpha_1 \alpha_2} [\bar{w}_{cd}(-\mathbf{q})]_{n_4 n_3}^{\alpha_4 \alpha_3} \rangle = \frac{4}{\sigma_{xx}} \delta_{ab;cd} \delta^{\alpha_1 \alpha_3} \delta^{\alpha_2 \alpha_4}$$

$$\times \delta_{n_{12}, n_{34}} D_q(\omega_{12}) \Big[\delta_{n_1 n_3} - \frac{32\pi T \Gamma_{ab}}{\sigma_{xx}} \delta^{\alpha_1 \alpha_2} D_q^{(ab)}(\omega_{12}) \Big], \tag{3.12}$$

where $\omega_{12} = \varepsilon_{n_1} - \varepsilon_{n_2} = 2\pi T n_{12} = 2\pi T (n_1 - n_2)$ and

$$D_q^{-1}(\omega_n) = q^2 + \frac{16 z \omega_n}{\sigma_{xx}},$$

$$[D_q^{(ab)}(\omega_n)]^{-1} = q^2 + \frac{16(z + \Gamma_{ab})\omega_n}{\sigma_{xx}}. \tag{3.13}$$

Here and further on we use notations $n_1, n_3, \ldots$ for nonnegative integers and $n_2, n_4, \ldots$ for negative ones.

Dynamical generalized susceptibility $\chi_{ab}(\omega, \mathbf{q})$ describes linear response to the space- and time-dependent parameter Δ_{ab}. It can be found from the following Matsubara susceptibility (Finkelstein, 1990)

$$\chi_{ab}(i\omega_n, \mathbf{q}) = \frac{2 z_{ab}}{\pi} - T z_{ab}^2 \langle\langle \mathrm{tr}\, I_n^\alpha t_{ab} Q(\mathbf{q})$$

$$\times \mathrm{tr}\, I_{-n}^\alpha t_{ab} Q(-\mathbf{q}) \rangle\rangle \tag{3.14}$$

with the help of analytic continuation to real frequencies: $i\omega_n \rightarrow \omega + i0^+$ (Finkelstein, 1984b). In the tree-level approximation Eq. 3.14 becomes

$$\chi_{ab}(i\omega_n, \mathbf{q}) = \frac{2 z_{ab}}{\pi} \left(1 - \frac{16 z_{ab} \omega_n}{\sigma_{xx}} D_q^{ab}(\omega_n) \right). \tag{3.15}$$

In some cases of matrix Γ_{ab} the action $\mathcal{S}_\sigma + \mathcal{S}_F$ can be invariant under global rotations $Q \rightarrow uQu^{-1}$ with a spatially independent matrix $u = u_{a_0 b_0} t_{a_0 b_0}$. Provided such invariance of the action exists, the quantity $\operatorname{tr} t_{a_0 b_0} Q$ is conserved. As a consequence, the corresponding Ward identity holds: $\chi_{a_0 b_0}(\omega, \mathbf{q} = 0) = 0$. Then Eq. 3.15 implies that $z_{a_0 b_0} = z + \Gamma_{a_0 b_0}$. In general case, there is no simple relation between z_{ab} and Γ_{ab}.

The symmetry-breaking term $\mathcal{S}_{\mathrm{SB}}$ with $\Delta_{a_0 b_0}$ changes Eq. 3.12. The propagator reads

$$
\langle [w_{ab}(\mathbf{q})]_{n_1 n_2}^{\alpha_1 \alpha_2} [w_{cd}(-\mathbf{q})]_{n_4 n_3}^{\alpha_4 \alpha_3} \rangle = \frac{4}{\sigma_{xx}} \delta^{\alpha_1 \alpha_3} \delta^{\alpha_2 \alpha_4} \delta_{n_{12}, n_{34}}
$$
$$
\times \left\{ \widehat{D}_q(\omega_{12}) \left[\delta_{n_1 n_3} - \frac{32\pi T}{\sigma_{xx}} \delta^{\alpha_1 \alpha_2} \widehat{\Gamma} \widehat{D}_q^{(\mathrm{int})}(\omega_{12}) \right] \right\}_{ab,cd}, \quad (3.16)
$$

where $\widehat{\Gamma} = \operatorname{diag}\{\Gamma_{00}, \Gamma_{01}, \Gamma_{02}, \Gamma_{03}, \Gamma_{10}, \ldots, \Gamma_{33}\}$ and

$$
\left[\widehat{D}_q(\omega_n) \right]_{ab,cd}^{-1} = D_q^{-1}(\omega_n) \delta_{ab,cd} + \frac{8i z_{a_0 b_0} \Delta_{a_0 b_0}}{\sigma_{xx}} C_{ab,cd}^{a_0 b_0},
$$
$$
\left[\widehat{D}_q^{(\mathrm{int})}(\omega_n) \right]_{ab,cd}^{-1} = \left[\widehat{D}_q(\omega_n) \right]_{ab,cd}^{-1} + \frac{16\Gamma_{ab}\omega_n}{\sigma_{xx}} \delta_{ab,cd}. \quad (3.17)
$$

Here $C_{cd;ef}^{ab}$ denotes the $SU(4)$ structure constants: $[t_{cd}, t_{ef}] = \sum_{ab} C_{cd;ef}^{ab} t_{ab}$. As one can find from Eq. 3.17, a part of diffusive modes becomes massive and does not lead to logarithmic divergencies at length scales $L \gg \sqrt{\sigma_{xx}/(z_{a_0 b_0} \Delta_{a_0 b_0})}$. One can check that it is the modes determining renormalization of the corresponding generalized susceptibility $\chi_{a_0 b_0}(\omega, \mathbf{q})$. Therefore, at length scales $L \gg \sqrt{\sigma_{xx}/(z_{a_0 b_0} \Delta_{a_0 b_0})}$ the physical observable $z'_{a_0 b_0}$ does not acquire renormalization.

3.2.3.2 One-loop renormalization of physical observables

At length scales $L \ll \min_{ab} \sqrt{\sigma_{xx}/(z_{ab} \Delta_{ab})}$ one can neglect the symmetry-breaking term $\mathcal{S}_{\mathrm{SB}}$. Using Eq. 3.12 we compute averages in Eq. 3.8 in the one-loop approximation. After analytic continuation we obtain

$$
\sigma'_{xx} = \sigma_{xx} + \frac{64}{\sigma_{xx} d} \Im \int \frac{d^d \mathbf{p}}{(2\pi)^d} \, p^2 \sum_{ab} \Gamma_{ab} \int d\omega
$$
$$
\times \frac{\partial}{\partial \omega} \left(\omega \coth \frac{\omega}{2T} \right) [D_p^R(\omega)]^2 D_p^{(ab),R}(\omega). \quad (3.18)
$$

Here $D_p^R(\omega)$ and $D_p^{(ab),R}(\omega)$ denote retarded propagators corresponding to Matsubara propagators $D_p(\omega_n)$ and $D_p^{(ab)}(\omega_n)$, respectively:

$$[D_p^R(\omega)]^{-1} = p^2 - \frac{16i\omega z}{\sigma_{xx}},$$

$$[D_p^{(ab),R}(\omega)]^{-1} = p^2 - \frac{16i\omega(z + \Gamma_{ab})}{\sigma_{xx}}. \tag{3.19}$$

To find the physical observable z', one has to evaluate the thermodynamic potential Ω. In the one-loop approximation we find

$$T^2 \frac{\partial \Omega/T}{\partial T} = 8N_r T \sum_{\omega_n > 0} \omega_n \Bigg[z + \frac{2}{\sigma_{xx}} \sum_{ab} \int \frac{d^d\mathbf{p}}{(2\pi)^d}$$

$$\times \Big((z + \Gamma_{ab}) D_p^{(ab)}(\omega_n) - z D_p(\omega_n) \Big) \Bigg]. \tag{3.20}$$

Hence, using Eq. 3.9, we obtain

$$z' = z + \frac{2}{\sigma_{xx}} \sum_{ab} \Gamma_{ab} \int \frac{d^d\mathbf{p}}{(2\pi)^d} D_p^R(0). \tag{3.21}$$

The physical observables z'_{ab} can be found from the generalized susceptibilities $\chi_{ab}(\omega, \mathbf{q})$ in the static limit $\omega = 0$ and $q \to 0$. In the one-loop approximation we find

$$z'_{ab} = z_{ab} + \frac{32\pi z_{ab}^2}{\sigma_{xx}^2} \sum_{cd;ef} [\mathcal{C}_{cd;ef}^{ab}]^2 \int \frac{d^d\mathbf{p}}{(2\pi)^d} T \sum_{\omega_m > 0}$$

$$\times \Big[D_p^{(ef)}(\omega_m) D_p^{(cd)}(\omega_m) - D_p^2(\omega_m) \Big]. \tag{3.22}$$

In general, even in the absence of the symmetry-breaking term S_{SB}, the observables z_{ab}, z and Γ_{ab} are unrelated. Therefore, renormalization of Γ_{ab} needs to be found by other means. For example, it can be done by the background field renormalization method (Amit, 1984). Then we obtain

$$\Gamma'_{ab} = \Gamma_{ab} - \int \frac{d^d\mathbf{p}}{(2\pi)^d} D_p(0) \sum_{cd;ef} \frac{\Gamma_{cd}}{8\sigma_{xx}} \big[\text{sp}(t_{cd} t_{ef} t_{ab}) \big]^2$$

$$- \frac{32\pi T}{\sigma_{xx}^2} \sum_{\omega_m > 0} \int \frac{d^d\mathbf{p}}{(2\pi)^d} \sum_{cd;ef} [\mathcal{C}_{cd;ef}^{ab}]^2 \Big\{ \Gamma_{ab}^2 D_p^2(\omega_m)$$

$$- \big[\Gamma_{cd} \Gamma_{ef} + \Gamma_{ab}^2 - 2\Gamma_{ab} \Gamma_{cd} \big] D_p^{(cd)}(\omega_m) D_p^{(ef)}(\omega_m) \Big\}. \tag{3.23}$$

Here sp denotes the trace over the spin and isospin spaces.

3.2.4 One-Loop RG Equations

As usual (see, for example, Amit, 1984) we derive the following one-loop RG equations in $d = 2$ from Eqs. 3.18–3.23 (Burmistrov et al., 2011):

$$\frac{d\sigma_{xx}}{dy} = -\frac{2}{\pi}\left[2 + \sum_{ab} f(\Gamma_{ab}/z)\right], \quad y = \ln L/l$$

$$\frac{d\Gamma_{ab}}{dy} = -\frac{1}{2\pi\sigma_{xx}}\sum_{cd;ef}\left[[\mathrm{sp}(t_{cd}t_{ef}t_{ab})]^2\frac{\Gamma_{cd}}{8} + [C_{cd;ef}^{ab}]^2 \right.$$

$$\left. \times \left(\frac{\Gamma_{ab}^2}{z} - \frac{(\Gamma_{ab} - \Gamma_{cd})(\Gamma_{ab} - \Gamma_{ef})}{\Gamma_{cd} - \Gamma_{ef}}\ln\frac{z+\Gamma_{cd}}{z+\Gamma_{ef}}\right)\right],$$

$$\frac{dz}{dy} = \frac{1}{\pi\sigma_{xx}}\sum_{ab}\Gamma_{ab}, \quad f(x) = 1 - \frac{1+x}{x}\ln(1+x). \qquad (3.24)$$

RG equations, Eq. 3.24, describe the behavior of the physical observables at length scales $l \ll L \ll \min_{ab}\sqrt{\sigma_{xx}/(z_{ab}\Delta_{ab})}$ at $T = 0$. They generalize previous results for a two-valley electron system (Finkelstein, 1990; Punnoose and Finkelstein, 2011) to the case of different interaction amplitudes Γ_{ab}. It is important to mention that the symmetric situation in which all Γ_{ab} except Γ_{00} are equal is unstable (see Appendix A.1). As we shall discuss further, different values of Γ_{ab} can be realized experimentally in electron systems with spin and isospin degrees of freedom. RG equations Eqs. 3.24 lead to a number of new effects as compared to the standard case (Finkelstein, 1990; Punnoose and Finkelstein, 2011): $\Gamma_{ab} = \Gamma$ for $(ab) \neq (00)$.

We note that we have added 2 into the square brackets of the right-hand side of the RG equation for the conductivity. This term describes the weak localization contribution due to cooperons. As is well known (Altshuler and Aronov, 1985; Altshuler et al., 1981; Altshuler and Aronov, 1981), this contribution is insensitive to symmetry-breaking terms S_{SB}.[j] We remind that the weak localization contribution is suppressed by a weak perpendicular

---

[j]To be precise, it is true in the absence of spin (or isospin) flips which results in appearance in the action terms $\mathrm{tr}[\Sigma, Q]^2$ where matrix Σ is determined by a particular (iso)spin flip mechanism.

magnetic field $B \gtrsim 1/D\tau_\phi$. Also cooperons contribute to the conductivity renormalization due to interaction in the Cooper channel (Finkelstein, 1990). However, in 2D electron systems the Cooper channel interaction is repulsive and renormalizes to zero. In addition, weak perpendicular magnetic field $B \gtrsim T/D$ suppresses such contributions (Altshuler et al., 1982).

Using Eq. 3.22 we find one-loop RG equations for the physical observables z_{ab} (Burmistrov et al., 2011):

$$\frac{dz_{ab}}{dy} = \frac{z_{ab}^2}{2\pi\sigma_{xx}} \sum_{cd;ef} \frac{\left[\mathcal{C}_{cd;ef}^{ab}\right]^2}{\Gamma_{cd} - \Gamma_{ef}} \left[\ln \frac{z + \Gamma_{cd}}{z + \Gamma_{ef}} - \frac{\Gamma_{cd} - \Gamma_{ef}}{z} \right].$$

$$(3.25)$$

As one can see from Eq. 3.24, the relation $z_{ab} = z + \Gamma_{ab}$ does not satisfy Eq. 3.25, generally.

RG equations, Eq. 3.24, describe the length-scale dependence of the physical observables at $T = 0$. At a finite temperature RG equations have to be stopped at the temperature-induced length scale L_{in}. In the case $\sigma_{xx} \gg 1$ and at a finite temperature, RG equations, Eq. 3.24, are valid upto the length scale $L_T = \sqrt{\sigma_{xx}/(zT)}$. At $L_T \ll L \ll L_\phi = \sqrt{\sigma_{xx}\tau_\phi/z}$ the conductivity is changed due to weak localization contribution only.

3.2.5 Conductivity Corrections due to Small Symmetry-Breaking Terms

The symmetry-breaking term Eq. 3.5 does not affect the RG equations at scales $L \ll \sqrt{\sigma_{xx}/(z_{ab}\Delta_{ab})}$ (or at temperatures $T \gg \Delta_{ab}$). However, they still change the temperature behavior of the physical observables. In the presence of the splitting Δ_{ab} the one-loop correction to the conductivity becomes

$$\sigma'_{xx} = \sigma_{xx} + \frac{64}{\sigma_{xx}d} \int \frac{d^d\mathbf{p}}{(2\pi)^d} \, p^2 \int d\omega \frac{\partial}{\partial\omega} \left(\omega \coth \frac{\omega}{2T} \right)$$
$$\times \Im \, \text{sp} \left([\hat{D}_p^R(\omega)]^2 \hat{\Gamma} \hat{D}_p^{(\text{int}),R}(\omega) \right). \qquad (3.26)$$

Hence, we find the following result for the second (the lowest) order in Δ_{ab} correction to the conductivity:

$$\delta\sigma'_{xx} = \frac{128 z_{ab}^2 \Delta_{ab}^2}{\sigma_{xx}^3 d} \int d\omega \frac{\partial}{\partial\omega} \left(\omega \coth\frac{\omega}{2T}\right) \sum_{cd} \Gamma_{cd}$$

$$\times \mathrm{sp}[t_{cd}, t_{ab}]^2 \Im \int \frac{d^d\mathbf{p}}{(2\pi)^d} \, p^2$$

$$\times \frac{\partial^2}{(\partial p^2)^2} \left([D_p^R(\omega)]^2 D_p^{(cd), R}(\omega)\right). \tag{3.27}$$

Integrating over momentum and frequency, we obtain

$$\delta\sigma'_{xx} = \frac{3\zeta(3)}{128\pi^3} \sum_{cd} \mathrm{sp}[t_{cd}, t_{ab}]^2 \frac{\gamma_{cd}}{1+\gamma_{cd}} \left(\frac{z_{ab}\Delta_{ab}}{zT}\right)^2. \tag{3.28}$$

Equation 3.28 generalizes the results (Castellani et al., 1998; Zala et al., 2001b; Altshuler and Aronov, 1985) for the correction to the magnetoresistance in a small parallel magnetic field to the case of arbitrary interaction amplitudes. The parameters z_{ab} and z, as well as the interaction amplitudes γ_{cd}, should be taken at the length scale $L_T = \sqrt{\sigma_{xx}/zT}$.

3.2.6 Dephasing Time

One of the important characteristics of interacting electron system is the dephasing time (Altshuler and Aronov, 1985). In particular, the temperature dependence of the dephasing time determines the T dependence of the weak localization contribution to the conductivity. Generalizing known results (Schmidt, 1974; Altshuler and Aronov, 1979b; Narozhny et al., 2002) to the case of electrons with spin and isospin degrees of freedom we find the total dephasing rate at $T \gg \Delta_{s,v}$ (Burmistrov et al., 2011):

$$\frac{1}{\tau_\phi} = -\frac{4}{\sigma_{xx}} \int_{\tau_\phi^{-1}} d\omega \int \frac{d^2q}{(2\pi)^2} \frac{1}{\sinh(\omega/T)} \Re D_q^R(\omega)$$

$$\times \Im \sum_{ab} \frac{\Gamma_{ab}}{z} D_q^{(ab), R}(\omega)[D_q^R(\omega)]^{-1}. \tag{3.29}$$

where

$$\mathcal{U}^{(ab)}(q, \omega) = \frac{\Gamma_{ab}}{z} D_q^{(ab), R}(\omega)[D_q^R(\omega)]^{-1}. \tag{3.30}$$

Integrating over momentum and frequency and then cutting off the logarithmic divergence in the infrared by the dephasing time we obtain for $1/\tau_\phi \gg \Delta_{s,v}$

$$\frac{1}{\tau_\phi} = \frac{T}{2\sigma_{xx}} \left(\sum_{ab} \frac{\gamma_{ab}^2}{2 + \gamma_{ab}} \right) \ln T \tau_\phi. \tag{3.31}$$

We mention that in Eq. 3.31 interaction amplitudes γ_{ab} and conductivity σ_{xx} corresponds to the length scale $L_T = \sqrt{\sigma_{xx}/zT}$.

3.3 Spin–Valley Interplay in a 2D Disordered Electron Liquid

3.3.1 Introduction

In this section we use general RG equations (Eq. 3.24) for description of the temperature dependence of resistance and spin/valley susceptibilities in 2D electron system in a Si-MOSFET. We assume that there is a parallel magnetic field B producing Zeeman splitting $\Delta_s = g_L \mu_B B \ll \tau_{tr}^{-1}$. Here g_L and μ_B denotes g-factor and the Bohr magneton, respectively. Also, we assume that a finite valley splitting Δ_v and intervalley scattering time τ_v exist. We consider the case $\tau_{so}^{-1}, \tau_v^{-1} \ll \Delta_v \ll \tau_{tr}^{-1}$ where τ_{so} stands for the spin relaxation time due to spin–orbit coupling. Experiments of (Kuntsevich et al., 2007; Klimov et al., 2008) indicate that such assumptions are reasonable for 2D electrons in a Si-MOSFET with electron concentrations near the metal–insulator transition.

3.3.2 Microscopic Hamiltonian

To describe a 2D two-valley disordered electron system realized in a Si(001)-MOSFET, it is convenient to write the electron annihilation operator with spin projection $\sigma/2$ on the z axis as follows (Brener et al., 2003; Nestoklon et al., 2006):

$$\psi_\sigma(\mathbf{R}) = \sum_{\tau=\pm} \psi_\tau^\sigma(\mathbf{r}) \varphi(z) [e^{izQ/2} + \tau e^{-izQ/2}]/\sqrt{2}. \tag{3.32}$$

Here the z axis is perpendicular to the (001) plane, $\mathbf{r}$ denotes 2D coordinate vector, and $\mathbf{R} = \mathbf{r} + z\mathbf{e_z}$. Subscript $\tau = \pm 1$

enumerates valley such that ψ_τ^σ denotes the annihilation operator for electron with z axis spin projection $\sigma/2$ and isospin projection $\tau/2$. We choose the envelope function $\varphi(z)$ to be normalized. In what follows, we neglect overlap $\int dz\, \varphi^2(z) \sin(Qz)$. Vector $\mathbf{Q} = (0, 0, Q)$ corresponds to the shortest distance between valleys in the reciprocal space. Its length can be estimated as $Q \approx a_{\text{lat}}^{-1}$ where a_{lat} stands for the lattice constant (Ando et al., 1982).

The 2D two-valley electron system is described by the following grand partition function:

$$Z = \int \mathcal{D}[\bar{\psi}, \psi] \exp S[\bar{\psi}, \psi], \tag{3.33}$$

where imaginary time action reads ($\beta = 1/T$)

$$S = -\int_0^\beta dt \left\{ \int d\mathbf{r}\, \bar{\psi}_\tau^\sigma(\mathbf{r}, t) \left[\partial_t + \mathcal{H}_0\right] \psi_\tau^\sigma(\mathbf{r}, t) - \mathcal{L}_{\text{dis}} - \mathcal{L}_{\text{int}} \right\}. \tag{3.34}$$

Single-particle Hamiltonian

$$\mathcal{H}_0 = -\frac{\nabla^2}{2m_e} - \mu + \frac{\Delta_s}{2}\sigma + \frac{\Delta_v}{2}\tau \tag{3.35}$$

describes 2D quasiparticle with mass m_e in the presence of a parallel magnetic field B ($\Delta_s = g_{LM} \mu_B B$) and valley splitting Δ_v. Here μ denotes the chemical potential. Lagrangian

$$\mathcal{L}_{\text{dis}} = -\int d\mathbf{r}\, \bar{\psi}_{\tau_1}^\sigma(\mathbf{r}) V_{\tau_1 \tau_2}(\mathbf{r}) \psi_{\tau_2}^\sigma(\mathbf{r}) \tag{3.36}$$

encodes scattering electrons off a random potential $V(\mathbf{R})$. Matrix elements of the random potential can be written as

$$V_{\tau_1 \tau_2}(\mathbf{r}) = \frac{1}{2} \int dz\, V(\mathbf{R}) \varphi^2(z) \left[1 + \tau_1 \tau_2 + \tau_1 e^{izQ} + \tau_2 e^{-izQ}\right]. \tag{3.37}$$

In general, matrix elements $V_{\tau_1 \tau_2}(\mathbf{r})$ produce not only intravalley scattering but also intervalley scattering. We assume that random potential $V(\mathbf{R})$ is Gaussian with $\langle V(\mathbf{R}) \rangle = 0$ and

$$\langle V(\mathbf{R}_1) V(\mathbf{R}_2) \rangle = W(|\mathbf{r}_1 - \mathbf{r}_2|, |z_1 - z_2|). \tag{3.38}$$

Here function W decays at the typical length scale d_W. As one can check, if

$$Q^{-1} \ll d_W, \quad \left[\int \varphi^4(z) dz\right]^{-1} \ll n_e^{-1/2}, \tag{3.39}$$

the intervalley scattering is negligible and

$$\langle V_{\tau_1\tau_2}(\mathbf{r}_1)V_{\tau_3\tau_4}(\mathbf{r}_2)\rangle = \frac{1}{2\pi\nu_\star\tau_i}\delta_{\tau_1\tau_2}\delta_{\tau_3\tau_4}\delta(\mathbf{r}_1-\mathbf{r}_2),$$

$$\frac{1}{\tau_i} = 2\pi\nu_\star\int d^2\mathbf{r}dz_1dz_2\ W(|\mathbf{r}|,|z_1-z_2|)\varphi^2(z_1)\varphi^2(z_2).$$

$$(3.40)$$

Under conditions Eq. 3.39, Lagrangian describing interaction is invariant under global $SU(4)$ rotations of electron operators ψ_τ^σ in spin and isospin spaces:

$$\mathcal{L}_{\text{int}} = -\frac{1}{2}\int d\mathbf{r}_1 d\mathbf{r}_2\ \bar{\psi}_{\tau_1}^{\sigma_1}(\mathbf{r}_1)\psi_{\tau_1}^{\sigma_1}(\mathbf{r}_1)U(|\mathbf{r}_1-\mathbf{r}_2|)\bar{\psi}_{\tau_2}^{\sigma_2}(\mathbf{r}_2)\psi_{\tau_2}^{\sigma_2}(\mathbf{r}_2).$$

$$(3.41)$$

Here $U(r) = e^2/\varepsilon r$ where ε stands for the dielectric constant. The low-energy part of $\mathcal{L}_{\text{int}}$ can be written as (see, for example, Zala et al., 2001a; Finkelstein, 1990; Belitz and Kirkpatrick, 1994)

$$\mathcal{L}_{\text{int}} = \frac{1}{4\nu_\star}\int_{q l\lesssim 1}\frac{d\mathbf{q}}{(2\pi)^2}\sum_{a,b=0}^{3}\mathbb{F}_{ab}(q)m^{ab}(\mathbf{q})m^{ab}(-\mathbf{q}),$$

$$m^{ab}(\mathbf{q}) = \int\frac{d\mathbf{k}}{(2\pi)^2}\bar{\Psi}(\mathbf{k}+\mathbf{q})t_{ab}\Psi(\mathbf{k}) \qquad (3.42)$$

with $\bar{\Psi} = \{\bar{\psi}_+^+, \bar{\psi}_+^-, \bar{\psi}_-^+, \bar{\psi}_-^-\}$, $\Psi = \{\psi_+^+, \psi_+^-, \psi_-^+, \psi_-^-\}^T$, and

$$\mathbb{F}(q) = \begin{pmatrix} F_s & F_t & F_t & F_t \\ F_t & F_t & F_t & F_t \\ F_t & F_t & F_t & F_t \\ F_t & F_t & F_t & F_t \end{pmatrix}. \qquad (3.43)$$

Here the quantity F_t is a standard Fermi liquid interaction parameter in the triplet channel. In the random phase approximation (RPA) it can be estimated as

$$F_t = -\frac{\nu_\star}{2}\langle U^{\text{scr}}(0)\rangle_{\text{FS}}, \qquad F_s = \nu_\star U(q) + F_t,$$

$$\langle U^{\text{scr}}(0)\rangle_{\text{FS}} = \int_0^{2\pi}\frac{d\theta}{2\pi}U^{\text{scr}}(2k_F\sin(\theta/2),0), \qquad (3.44)$$

where dynamically screened Coulomb interaction is as follows

$$U^{\text{scr}}(q,\omega) = \frac{U(q)}{1+U(q)\Pi(q,\omega)},$$

$$\Pi(q,\omega) = \frac{\nu_\star Dq^2}{Dq^2-i\omega}. \qquad (3.45)$$

We mention that k_F denotes the Fermi momentum of electrons in a single valley. The quantity F_s involves Coulomb interaction at a small momentum. In the limit $q \to 0$, $F_s(q) \approx \varkappa/q \to \infty$ where $\varkappa = 2\pi e^2 \nu_*/\varepsilon$ stands for the inverse static screening length. We remind known results for the interaction parameter F_t (Zala et al., 2001a):

$$F_t = -\int_0^{2\pi} \frac{d\theta}{4\pi} \frac{\varkappa}{2k_F \sin(\theta/2) + \varkappa} = -\frac{1}{2\pi} \mathcal{G}_0(\varkappa/2k_F),$$

$$\mathcal{G}_0(x) = \frac{x}{\sqrt{1-x^2}} \ln \frac{1 + \sqrt{1-x^2}}{1 - \sqrt{1-x^2}}. \qquad (3.46)$$

Under condition $\varkappa/k_F \ll 1$ which justifies RPA we obtain $\mathcal{G}_0(x) \approx x \ln(2/x)$.

3.3.3 *SU*(4) Symmetric Case

It is convenient to introduce interaction $\gamma_t = -F_t/(1 + F_t)$. Then for all interaction amplitudes Γ_{ab} except Γ_{00} the following relations hold: $\Gamma_{ab} = z\gamma_t$ for $(ab) \neq (00)$, and $\Gamma_{00} = -z$. Therefore, at high energies $|E| \approx 1/\tau_{tr}$ and short length scales $L \approx l$ electron–electron interaction in a Si-MOSFET does not discriminate inter- and intravalley interactions. The presence in the Hamiltonian Eq. 3.35 of spin Δ_s and valley Δ_v splittings leads to the symmetry-breaking terms, Eq. 3.5, in action, Eq. 3.1. At short length scales $L \ll L_{s,v} = \sqrt{\sigma_{xx}/(16 z_{s,v} \Delta_{s,v})}$ where $z_{s,v} = z(1 + \gamma_t)$ or, equivalently, at high temperatures $T \gg \Delta_{s,v}$, the $SU(4)$ symmetry-breaking term Eq. 3.5 is not important.

Using general one-loop RG equations (Eq. 3.24), we obtain the well-known results for the 2D two-valley electron system (Punnoose and Finkelstein, 2001):

$$\frac{d\sigma_{xx}}{dy} = -\frac{2}{\pi}[2 + 1 + 15 f(\gamma_t)], \qquad (3.47)$$

$$\frac{d\gamma_t}{dy} = \frac{(1 + \gamma_t)^2}{\pi \sigma_{xx}}, \qquad (3.48)$$

$$\frac{d \ln z}{dy} = \frac{15\gamma_t - 1}{\pi \sigma_{xx}}. \qquad (3.49)$$

Solution of RG equations (Eqs. 3.47–3.48) yields metallic-type dependence of resistance $\rho = 1/\pi\sigma_{xx}$ at large y (ρ decreases with increase of y). We mention that dependence $\rho(y)$ has the maximum at value of y_{max} such that $\gamma_t(y_{max}) \approx 0.46$ (Punnoose and Finkelstein, 2001). The interaction amplitude γ_t increases monotonically with y.

3.3.4 *SU(2) × SU(2)* Case

Let us assume that $\Delta_s \gg \Delta_v$. Then it is possible to consider the intermediate length scales $L_s \ll L \ll L_v$ for which the symmetry-breaking term Eq. 3.5 with $\Delta_{03} = \Delta_s$ becomes important. If one use decomposition $Q = \sum_{ab} t_{ab} Q_{ab}$ then Eq. 3.16 yields that modes Q_{ab} with $b = 1, 2$ become massive and do not lead to logarithmic divergencies at $L_s \ll L \ll L_v$. Therefore, at such length scales the NLσM action is given by Eqs. 3.2 and 3.3 with

$$Q = \sum_{a=0}^{3} \sum_{b=0,3} t_{ab} Q_{ab}. \tag{3.50}$$

We mention that if one defines matrix fields $Q_\pm = (t_{00} \pm t_{03})Q/2$ then they will be decoupled in the absence of electron–electron interaction.

In the presence of Zeeman splitting one can distinguish interaction between electrons with equal spin projections and with different ones. It leads to the following form of matrix Γ:

$$\Gamma = \begin{pmatrix} \Gamma_s & 0 & 0 & \tilde{\Gamma}_t \\ \Gamma_t & 0 & 0 & \Gamma_t \\ \Gamma_t & 0 & 0 & \Gamma_t \\ \Gamma_t & 0 & 0 & \Gamma_t \end{pmatrix}. \tag{3.51}$$

Here combination $(\Gamma_s \pm \tilde{\Gamma}_t)$ describes interaction of electrons with equal/opposite spin projections.

Invariance of the action Eq. 3.3 with respect to global rotation Eq. 3.6 with $\chi_{ab} = \chi \delta_{a0}\delta_{b0}$ is guaranteed by the condition $\Gamma_s = -z$. Conservation of the total isospin yields $z_v = z + \Gamma_t$. Conservation of z component of the total spin (S_z) results in $z_s = z + \tilde{\Gamma}_t$. Since renormalization of the spin susceptibility, that is, z_s, is only possible due to interaction of diffusive modes with $S_z = \pm 1$ which are massive, there is no logarithmic divergences in z_s at $L_s \ll L \ll L_v$. Therefore, we obtain

$$\frac{dz_s}{dy} = \frac{d(\Gamma_s - \tilde{\Gamma}_t)}{dy} = 0. \qquad (3.52)$$

We emphasize that observables z_v and z_s which behave in the same way at small length scales $L \ll L_s \ll L_v$, have to flow differently at $L_s \ll L \ll L_v$. This is the reason for the appearance of interaction amplitude $\tilde{\Gamma}_t$ which has a different RG behavior in comparison with Γ_t.

Using general results, Eq. 3.24, we find the one-loop RG equations for the intermediate length scales $L_s \ll L \ll L_v$ ($\tilde{\gamma}_t = \tilde{\Gamma}_t/z$) (Burmistrov and Chtchelkatchev, 2008):

$$\frac{d\sigma_{xx}}{dy} = -\frac{2}{\pi}[2 + 1 + 6f(\gamma_t) + f(\tilde{\gamma}_t)], \qquad (3.53)$$

$$\frac{d\gamma_t}{dy} = \frac{1 + \gamma_t}{\pi \sigma_{xx}}(1 + 2\gamma_t - \tilde{\gamma}_t), \qquad (3.54)$$

$$\frac{d\tilde{\gamma}_t}{dy} = \frac{1 + \tilde{\gamma}_t}{\pi \sigma_{xx}}(1 - 6\gamma_t - \tilde{\gamma}_t), \qquad (3.55)$$

$$\frac{d\ln z}{dy} = -\frac{1}{\pi \sigma_{xx}}(1 - 6\gamma_t - \tilde{\gamma}_t). \qquad (3.56)$$

Here $y = \ln L/l_s$ where l_s is of the order of L_s.[k] In what follows we assume that $\tilde{\gamma}_t(0) = \gamma_t(0)$.

Figure 3.3 illustrates RG flow in coordinates $(\tilde{\gamma}_t, \gamma_t)$. There is unstable fixed point at $\tilde{\gamma}_t = 1$ and $\gamma_t = 0$. However, in the two-valley electron system this fixed point is inaccessible since RG flow starts near $\gamma_t = \tilde{\gamma}_t > 0$. As shown in Fig. 3.4, two different regimes for behavior of the resistance ρ are possible. Resistance decreases monotonously along curve a (see Fig. 3.3) which does not intersect curve d corresponding to equation $2 + 1 + 6f(\gamma_t) + f(\tilde{\gamma}_t) = 0$. If one flows along the curves b and c which intersect curve d, then resistance has the maximum. We mention that resistance behavior at large values of y is of metallic type. This is because the interaction amplitude γ_t grows at large y whereas $\tilde{\gamma}_t$ tends to -1. Therefore,

---

[k]To find exact relation between l_s and L_s one needs to solve a complicated crossover problem (see, for example, Amit, 1984) for description of behavior σ_{xx}, γ_t, $\tilde{\gamma}_t$ and z at length scales $L \approx L_s$.

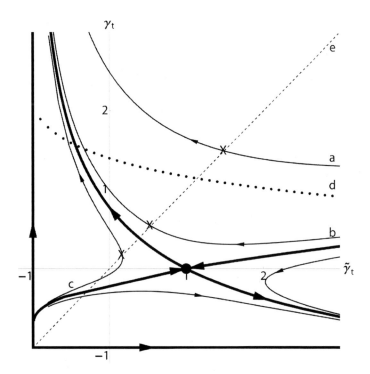

Figure 3.3 Projection of three-parameter $(\sigma_{xx}, \tilde{\gamma}_t, \gamma_t)$ RG flow Eqs. 3.53–3.55 to the plain $(\tilde{\gamma}_t, \gamma_t)$. The dotted curve d is determined by the equation $2 + 1 + 6f(\gamma_t) + f(\tilde{\gamma}_t) = 0$. The dashed line e is described by the equation $\gamma_t = \tilde{\gamma}_t$.

electrons with different spin projections become independent and Eqs. 3.53–3.56 transforms into RG equations for two independent copies of the single-valley system. It is well known (Finkelstein, 1990) that in the case of the single-valley electron system, RG equations yield metallic behavior of resistance at large values of y.

3.3.5 Completely Symmetry-Broken Case

At large length scales $L \gg L_v \gg L_s$ the symmetry-breaking term Eq. 3.5 with $\Delta_{30} = \Delta_v$ becomes important also. We remind that there is eight massless modes (Q_{ab} with $a = 0, 1, 2, 3$ and $b = 0, 3$) at intermediate length scales $L_s \ll L \ll L_v$. At large length scales $L \gg L_v \gg L_s$ only four modes Q_{00}, Q_{03}, Q_{30} and Q_{33} remain massless.

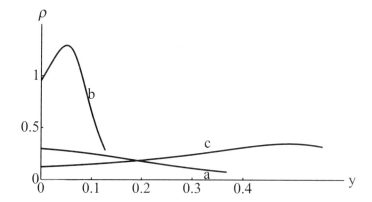

Figure 3.4 Schematic dependence of resistivity $\rho = 1/(\pi \sigma_{xx})$ on y. Initial values for curves a, b, and c correspond to crossing points X of curves a, b, and c with line e in Fig. 3.3.

Therefore, the NLσM action at $L \gg L_v \gg L_s$ is given by Eqs. 3.2 and 3.3 with matrix $Q = \sum_{a,b=0,3} t_{ab} Q_{ab}$. We mention that in the absence of electron–electron interaction four matrices $Q_s' = (t_{00} + st_{03} + s't_{30} + ss't_{33})Q/4$ where $s, s' = \pm$ are decoupled. In the presence of strong spin and valley splitting one can distinguished interaction between electrons with equal and opposite spin and isospin projections. Hence, the matrix Γ acquires the following general form:

$$\Gamma = \begin{pmatrix} \Gamma_s & 0 & 0 & \tilde{\Gamma}_t \\ 0 & 0 & 0 & 0 \\ 0 & 0 & 0 & 0 \\ \Gamma_t & 0 & 0 & \hat{\Gamma}_t \end{pmatrix}. \tag{3.57}$$

Here $(\Gamma_s + s\tilde{\Gamma}_t + s'\Gamma_t + ss'\hat{\Gamma}_t)/4$ corresponds to interaction of electrons with spin projection s and isospin projection s'. Invariance of the NLσM action under global rotation Eq. 3.6 with $\chi_{ab} = \chi \delta_{a0} \delta_{b0}$ implies that the following condition holds: $\Gamma_s = -z$. Conservation of the z-th isospin component yields relation $z_v = z + \Gamma_t$, which holds also at the intermediate length scales $L_s \ll L \ll L_v$. Conservation of the z-th component of the spin results in the relation $z_s = z + \tilde{\Gamma}_t$. In addition to spin and valley susceptibility, one can define spin–valley susceptibility which describes the linear response to the Δ_{33} splitting. In the static limit it is determined by the quantity $z_{sv} =$

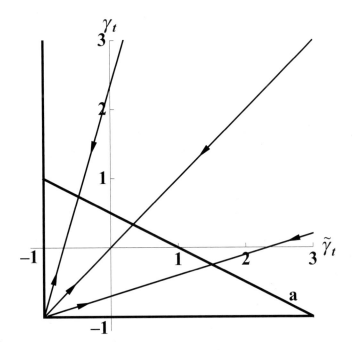

Figure 3.5 Projection of three-parameter $(\sigma_{xx}, \tilde{\gamma}_t, \gamma_t)$ RG flow Eqs. 3.59–3.61 to the plain $(\tilde{\gamma}_t, \gamma_t)$. Line a corresponds to the equation $2\gamma_t + \tilde{\gamma}_t = 1$.

$z + \hat{\Gamma}_t$. At length scales $L \gg L_v \gg L_s$ spin, valley and spin-valley susceptibilities do not renormalize. Therefore, we obtain the following RG equations:

$$\frac{dz_v}{dy} = \frac{d(\Gamma_t - \Gamma_s)}{dy} = 0, \quad \frac{dz_s}{dy} = \frac{d(\tilde{\Gamma}_t - \Gamma_s)}{dy} = 0,$$

$$\frac{dz_{sv}}{dy} = \frac{d(\hat{\Gamma}_t - \Gamma_s)}{dy} = 0. \tag{3.58}$$

Since interaction amplitudes $\hat{\Gamma}_t$ and Γ_t coincide at length scales $L \approx L_v$, and at larger scales they are renormalized in the same way, we consider them as equal: $\hat{\Gamma}_t \equiv \Gamma_t$.

Using general RG equations (Eq. 3.24), we obtain the RG equation for length scales $L \gg L_v \gg L_s$ (Burmistrov and Chtchelkatchev, 2008):

$$\frac{d\sigma_{xx}}{dy} = -\frac{2}{\pi}[2 + 1 + 2f(\gamma_t) + f(\tilde{\gamma}_t)], \qquad (3.59)$$

$$\frac{d\gamma_t}{dy} = \frac{1 + \gamma_t}{\pi\sigma_{xx}}(1 - 2\gamma_t - \tilde{\gamma}_t), \qquad (3.60)$$

$$\frac{d\tilde{\gamma}_t}{dy} = \frac{1 + \tilde{\gamma}_t}{\pi\sigma_{xx}}(1 - 2\gamma_t - \tilde{\gamma}_t), \qquad (3.61)$$

$$\frac{d\ln z}{dy} = -\frac{1}{\pi\sigma_{xx}}[1 - 2\gamma_t - \tilde{\gamma}_t]. \qquad (3.62)$$

Here $y = \ln L/l_v$ where the scale l_v is of the order of L_v.

RG flow for Eqs. 3.59–3.61 in $(\gamma_t, \tilde{\gamma}_t)$ plane is shown in Fig. 3.4. There is the line of fixed points determined by equation $2\gamma_t + \tilde{\gamma}_t = 1$. RG flow in $(\gamma_t, \tilde{\gamma}_t)$ plane is just the straight lines $(1 + \tilde{\gamma}_t)/(1 + \gamma_t) = $ const. The curve described by equation $2 + 1 + 2f(\gamma_t) + f(\tilde{\gamma}_t) = 0$ lies in the region of relatively large values of γ_t and $\tilde{\gamma}_t$. Therefore, if initial values $\gamma_t(0)$ and $\tilde{\gamma}_t(0)$ are not large resistance $\rho(y)$ will be a monotonous increasing function of y, that is, it will demonstrate insulating behavior.

3.3.6 Discussion and Comparison with Experiments

Keeping in mind discussion at the end of Section 3.2.4 we assume that RG equations (Eqs. 3.47–3.48, 3.53–3.55, and 3.59–3.61) describe temperature dependence of the physical observables as listed in Table 3.1.

Let us start from the case of zero Zeeman splitting, $\Delta_s = 0$. We assume that the following condition holds $\Delta_v < T_{max}^{(I)}$. Here $T_{max}^{(I)}$ denotes the temperature at which resistance reaches the maximum

Table 3.1 Temperature regions in which RG equations are applied

RG equations	$\Delta_s = 0$	$\Delta_s \gg \Delta_v$
Eqs. 3.47–3.48	$\Delta_s \ll T \ll 1/\tau_{tr}$	$\Delta_s \ll T \ll 1/\tau_{tr}$
Eqs. 3.53–3.55	$1/\tau_v \ll T \ll \Delta_v$	$\Delta_v \ll T \ll \Delta_s$
Eqs. 3.59–3.61		$1/\tau_v \ll T \ll \Delta_v$

in accordance with Eqs. 3.47–3.48.[1] Then there are possible two type of behavior of resistance with temperature depending on the initial conditions (values of σ_{xx} and γ_t at temperature of the order of $1/\tau$). Both types of behavior are illustrated in Fig. 3.6a. Curve a is typical for $\rho(T)$ dependence observed in experiments on a Si-MOSFET (Kravchenko et al., 1994). Remarkably there exists more complicated behavior: $\rho(T)$ has *two* maximums (curve b in Fig. 3.6a). We mention that such $\rho(T)$ dependence has not been observed experimentally yet. We emphasize that in the absence of Zeeman splitting metallic behavior of $\rho(T)$ at temperatures $1/\tau_v \ll T \ll 1/\tau$ does not spoiled by the presence of finite valley splitting $\Delta_v \gg 1/\tau_v$. This result is in agreement with experimental observations on 2D two-valley electron systems (Simonian et al., 1997; Vitkalov et al., 2003; Pudalov et al., 2003; Gunawan et al., 2006, 2007).

In the presence of a weak parallel magnetic field $\Delta_s < T_{\max}^{(I)}$, three types of $\rho(T)$ behavior are possible as shown in Fig. 3.6b. In all three cases resistance has the maximum at temperature $T = T_{\max}^{(I)}$ and increases as temperature decreases. As one can see in Fig. 3.4, for temperatures $\Delta_s \lesssim T \lesssim \Delta_v$ three different scenarios for $\rho(T)$ are possible: metallic (curve a in Fig. 3.6b), insulating (curve b in Fig. 3.6b), and nonmonotonous (curve c in Fig. 3.6b). In the latter case, resistance can develop *two maximums* even in the presence of a parallel magnetic field. At lower temperatures $1/\tau_v \ll T \lesssim \Delta_v$ resistance increases with decrease of T. Therefore, the presence of both valley and spin splittings ($\Delta_s < \Delta_v$) yields the change of T dependence of resistance from metallic to insulating at low temperatures in agreement with experimental findings of Gunawan et al., Simonian et al., and Vitkalov et al. (Simonian et al., 1997; Vitkalov et al., 2003; Pudalov et al., 2003; Gunawan et al., 2006, 2007). We note that the typical temperature of metal-to-insulator crossover is of the order of either Δ_v or Δ_s depending on the initial values of σ_{xx} and γ_t at temperature of the order of $1/\tau_{\rm tr}$.

In the presence of a strong parallel magnetic field such that $\Delta_s > T_{\max}^{(I)}$ the maximum in $\rho(T)$ is absent and only two types of resistance behavior occur as schematically illustrated in Fig. 3.6c.

---

[1]For example, in a Si-MOSFET sample with a 2D electron concentration of about 10^{11} cm^{-2} the temperature $T_{\max}^{(I)}$ is equal to a few Kelvins.

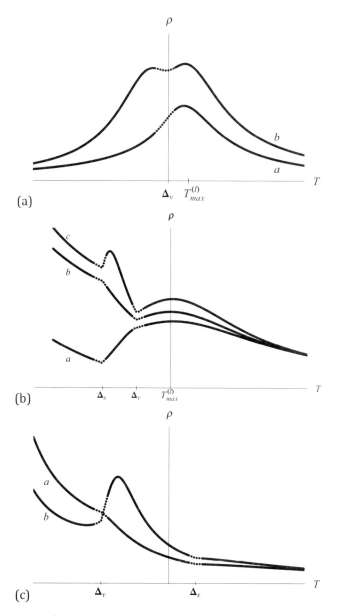

Figure 3.6 Schematic dependence of resistivity $\rho(T)$ for (a) zero spin splitting, (b) the case $\Delta_s < \Delta_v$, and (c) strong parallel magnetic field: $\Delta_v, T_{max}^{(I)} < \Delta_s$.

If the temperature $T_{max}^{(II)}$ of the resistance maximum found from RG equations (Eqs. 3.53–3.55) is such that $T_{max}^{(II)} < \Delta_v$, then $\rho(T)$ is monotonously increasing with lowering temperature (curve a in Fig. 3.6c). In the opposite case $T_{max}^{(II)} > \Delta_v$, the $\rho(T)$ dependence is shown in Fig. 3.6c by curve b. Therefore provided $\Delta_v > T_{max}^{(II)}$ the $\rho(T)$ dependence changes from metallic to insulating. It is in agreement with experimental data on magnetoresistance in a Si-MOSFET (Simonian et al., 1997; Knyazev et al., 2007). However, if valley splitting $\Delta_v < T_{max}^{(II)}$ then the maximum in $\rho(T)$ remains even in the presence of a parallel magnetic field.

In addition to interesting new temperature behavior of resistivity, RG equations derived above yield interesting predictions for T dependence of the spin and valley susceptibilities. Let us consider the ratio of static valley and spin susceptibilities χ_v/χ_s. A schematic dependence of χ_v/χ_s on T is shown in Fig. 3.7 at fixed values of valley splitting and different values of spin splitting. At high temperatures $T \gg \Delta_v, \Delta_s$ the ratio is equal to unity, $\chi_v/\chi_s = 1$. At $T \ll \Delta_v, \Delta_s$, we obtain

$$\frac{\chi_v}{\chi_s} \begin{cases} < 1 & , \quad \Delta_s < \Delta_v, \\ = 1 & , \quad \Delta_s = \Delta_v, \\ > 1 & , \quad \Delta_s > \Delta_v. \end{cases} \tag{3.63}$$

Therefore, the ratio χ_v/χ_s in the limit $T \to 0$ is sensitive to the ratio Δ_v/Δ_s.

To detect the second maximum in $\rho(T)$ experimentally one needs a system with wide interval between $1/\tau_v$ and Δ_v. Experimental data on magnetoresistance in a Si-MOSFET (Kuntsevich et al., 2007) allow to estimate intervalley scattering rate to be about 0.36 K for electron concentrations in the range of 3–6 $\times 10^{11}$ cm^{-2}. In the same range of concentrations the valley splitting varies only weakly. However, there is significant sample-to-sample variation of the valley splitting (Klimov et al., 2008). It was estimated to be in the range of 0.4–0.7 K. Due to such values of $1/\tau_v$ and Δ_v it is complicated to find the second maximum in $\rho(T)$ experimentally (Punnoose et al., 2010). We note that experiments in a Si-MOSFET on magnetoresistance in a parallel magnetic field demonstrates absence of dependence on relative orientation of current and magnetic field (Pudalov et al., 2003). It signifies smallness of effects due to spin–orbit coupling.

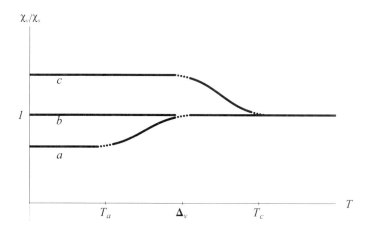

Figure 3.7 Schematic dependence of the ratio χ_v/χ_s on temperature in the cases (i) $\Delta_s < \Delta_v$ (curve a), (ii) $\Delta_s = \Delta_v$ (curve b), and (iii) $\Delta_s > \Delta_v$ (curve c). Characteristic temperature scales $T_{a,c} \equiv \Delta_s$.

The influence of the spin and valley splittings on resistivity discussed above was due to change in RG equations corresponding to temperatures $T \ll \Delta_{s,v}$. In the opposite case of small spin and valley splittings, $\Delta_{s,v} \ll T$ there are corrections to resistivity proportional to Δ_s^2 and Δ_v^2 (see Eq. 3.28). However, we mention that there is another source for the influence of $\Delta_{s,v} \ll T$ on temperature dependence of resistivity. It is due to possible dependence of initial (for RG flow) values of conductivity and interaction amplitudes on $\Delta_{s,v}$ which come from ballistic scales $T \gg 1/\tau_{tr}$ (Knyazev et al., 2007). This mechanism accounts for the temperature dependence of magnetoresistivity in a weak parallel magnetic field in experiments of Knyazev et al. (2007).

Finally, we reiterate the main result of the analysis presented in this section. The 2D two-valley disordered electron system realized in a Si-MOSFET in the presence of spin and/or valley splittings has to be described by three-parameter RG equations at length scales $L \gg \min\{L_s, L_v\}$. Recently, RG equations (Eqs. 3.53–3.55 and 3.59–3.61) were rederived by Punnoose (Punnoose, 2010a, 2010b). Also in these papers the behavior of the resistance at $T \ll 1/\tau_v \ll \Delta_v$ has been investigated. It was found that in the absence of Zeeman splitting RG equations at $T \ll 1/\tau_v \ll \Delta_v$ are exactly the same as in the case of the single-valley system, that is, metallic behavior of $\rho(T)$ should persist down to zero temperatures (Finkelstein, 1990). In the case of finite spin splitting $\Delta_s \gg \Delta_v \gg 1/\tau_v$, at $T \ll 1/\tau_v$ RG equations coincide with ones for the single-valley system in the presence of spin splitting $\Delta_s \gg T$. As is well known, under such circumstances the $\rho(T)$ dependence is of the insulating type (Finkelstein, 1990). Therefore, one can conclude that as temperature crosses the scale $1/\tau_v$ the resistance does not change a type of its temperature dependence (metallic or insulating).

3.4 2D Disordered Electron Liquid in the Double-Quantum-Well Heterostructure

3.4.1 Introduction

In this section we apply general RG equations (Eq. 3.24) to study of 2D electron system in a symmetric double quantum well with common disorder. We consider the case of equal electron concentrations and mobilities in both quantum wells. We shall assume that the following inequalities hold $1/\tau_{+-}, \Delta_s, \Delta_{SAS} \ll T \ll 1/\tau_{tr}$. Here Δ_{SAS} and $1/\tau_{+-}$ denote splitting and elastic scattering rate for transitions between symmetric and antisymmetric states in the double-quantum-well heterostructure, respectively. We note that such conditions correspond to the experiments of Minkov et al. (2010, 2011).

3.4.2 Microscopic Hamiltonian

In the case of symmetric double quantum well it is convenient to chose the following basis for electron annihilation operator:

$$\psi^\sigma(\mathbf{R}) = \psi_\tau^\sigma(\mathbf{r})\varphi_\tau(z), \quad \varphi_\tau(z) = \frac{\varphi_l(z) + \tau\varphi_r(z)}{\sqrt{2}}. \qquad (3.64)$$

Here we assume that motion of electrons along the z axis is confined due to quantum well barriers. Vector $\mathbf{r}$ denotes coordinates in the plane perpendicular to the z axis, and $\mathbf{R} = \mathbf{r} + z\mathbf{e_z}$. Superscript $\sigma = \pm$ stands for the spin projection to the z axis, and subscript $\tau = \pm$ enumerates symmetric $(+)$ and antisymmetric $(-)$ states.

Normalized wave functions $\varphi_{l,r}(z) = \varphi(z \pm d/2)$ describe an electron localized in the left/right well. We neglect their overlap. Also we assume that both quantum wells are narrow such that

$$\left[\int dz \, \varphi^4(z) \right]^{-1} \ll d \qquad (3.65)$$

where d is the distance between the centers of quantum wells.

In terms of electron operators $\psi_\tau^\sigma(\mathbf{r})$ and $\bar{\psi}_\tau^\sigma(\mathbf{r})$ the system is described by the grand partition function Eq. 3.33 with the action which has the same form as Eq. 3.34. The single-particle Hamiltonian is given by Eq. 3.35 in which now Δ_{SAS} plays a role of Δ_v. We mention that Δ_{SAS} can be estimated as $2\varphi(d/2)\varphi'(d/2)/m_e$ (Landau and Lifshitz, 1991). Scattering electrons off the random potential $V(\mathbf{R})$ is described by the Lagrangian $\mathcal{L}_{dis}$ given by Eq. 3.36. However, now the matrix elements $V_{\tau_1 \tau_2}(\mathbf{r})$ are defined as

$$V_{\tau_1 \tau_2}(\mathbf{r}) = \int dz \, V(\mathbf{R})\varphi_{\tau_1}(z)\varphi_{\tau_2}(z). \qquad (3.66)$$

In general, matrix elements $V_{\tau_1 \tau_2}$ produce transitions between symmetric and antisymmetric states in the double-quantum-well heterostructure. We note that in the case of mirror symmetry: $V(\mathbf{r}, z) = V(\mathbf{r}, -z)$, the transitions between symmetric and antisymmetric states are absent. We assume that impurities that create the random potential $V(\mathbf{R})$ are situated in the plane $z = 0$ which is in the middle between the quantum wells. This assumptions correspond to the double-quantum-well heterostructures studied by Minkov et al. (2010, 2011). Also we assume that $V(\mathbf{R})$ is Gaussian with a zero mean value and a two-point correlation function given by Eq. 3.38. Since for narrow quantum wells Eq. 3.65 holds, we can neglect the difference (of the order of $\varphi(d/2)\varphi'(d/2)$) in scattering rates between only symmetric or only antisymmetric states. Then in the case of random potential with short-range correlations[m] we find

$$\langle V_{\tau_1 \tau_2}(\mathbf{r}_1) V_{\tau_3 \tau_4}(\mathbf{r}_2) \rangle = \frac{1}{2\pi \nu_\star \tau_i} \delta_{\tau_1 \tau_2} \delta_{\tau_3 \tau_4} \delta(\mathbf{r}_1 - \mathbf{r}_2),$$

$$\frac{1}{\tau_i} = 2\pi \nu_\star \int d^2\mathbf{r} \, W(|\mathbf{r}|, d/2, d/2). \qquad (3.67)$$

---

[m] In the experiments of Minkov et al. (2010, 2011) random potential has been created by charged impurities situated near the plane $z = 0$ (in the middle between the quantum wells). In this case the typical range for the random potential is of the order of 3D screening length, $d_W \approx 1/\sqrt{\varkappa k_F}$. Provided the condition $d_W \ll l$ (equivalently, $k_F l \gg \sqrt{k_F/\varkappa}$) holds we can consider the random potential as short-range-correlated.

A small asymmetry in impurity distribution with respect to the z axis results in scattering between symmetric and antisymmetric states. Its rate can be estimated as $1/\tau_{+-} \approx b^2/(d^2\tau_i) \ll 1/\tau_i$ where b denotes the typical length scale characterizing asymmetry. In what follows we assume that temperature $T \gg 1/\tau_{+-}$ and, therefore, we will neglect such scattering.

The interacting Lagrangian is as follows

$$\mathcal{L}_{\text{int}} = -\frac{1}{2}\int d\mathbf{R}d\mathbf{R}'\rho(\mathbf{R}t)\,U\left(|\mathbf{R}-\mathbf{R}'|\right)\rho(\mathbf{R}'t), \tag{3.68}$$

where $U\left(\mathbf{R}\right) = e^2/\varepsilon R$ and the electron density operator $\rho(\mathbf{R}t) = \bar\psi_{\tau_1}^\sigma(\mathbf{r}t)\psi_{\tau_2}^\sigma(\mathbf{r}t)\varphi_{\tau_1}(z)\varphi_{\tau_2}(z)$. In the case of narrow quantum wells, Eq. 3.68 becomes

$$\mathcal{L}_{\text{int}} = -\frac{1}{8}\int d\mathbf{r}d\mathbf{r}'\ \bar\psi_{\tau_1}^{\sigma_1}(\mathbf{r}t)\psi_{\tau_2}^{\sigma_1}(\mathbf{r}t)\bar\psi_{\tau_3}^{\sigma_2}(\mathbf{r}'t)\psi_{\tau_4}^{\sigma_2}(\mathbf{r}'t)$$
$$\times\Big[(1+\tau_1\tau_2\tau_3\tau_4)U_{11}(|\mathbf{r}-\mathbf{r}'|)+$$
$$+(\tau_1\tau_2+\tau_3\tau_4)U_{12}(|\mathbf{r}-\mathbf{r}'|)\Big]. \tag{3.69}$$

Here

$$U_{11}(r) = \frac{e^2}{\varepsilon}\int dzdz'\,\frac{\varphi_l^2(z)\varphi_l^2(z')}{\sqrt{r^2+(z-z')^2}} \approx \frac{e^2}{\varepsilon r} \tag{3.70}$$

describes electron–electron interaction inside a quantum well. Interaction of electrons from different quantum wells is given by

$$U_{12}(r) = \frac{e^2}{\varepsilon}\int dzdz'\,\frac{\varphi_l^2(z)\varphi_r^2(z')}{\sqrt{r^2+(z-z')^2}} \approx \frac{e^2}{\varepsilon\sqrt{r^2+d^2}}. \tag{3.71}$$

Due to the difference between U_{11} and U_{12} Lagrangian $\mathcal{L}_{\text{int}}$ is not invariant under global $SU\left(4\right)$ rotation of ψ_τ^σ in the spin and isospin spaces. It is the form of $\mathcal{L}_{\text{int}}$ that makes the cases of electrons in the double-quantum-well heterostructure and a Si-MOSFET to be different. The low-energy part of $\mathcal{L}_{\text{int}}$ acquires the form of Eq. 3.42 with the following 4×4 matrix of the interaction parameters:

$$F(q) = \begin{pmatrix} F_s & F_t & F_t & F_t \\ \tilde{F}_s & F_t & F_t & F_t \\ F_v & F_v & F_v & F_v \\ F_v & F_v & F_v & F_v \end{pmatrix}. \tag{3.72}$$

In RPA the interaction parameters can be estimated as

$$F_t = -\frac{\nu_\star}{2} \langle U_{11}^{\text{scr}}(0) \rangle_{\text{FS}}, \quad F_v = -\frac{\nu_\star}{2} \langle U_{12}^{\text{scr}}(0) \rangle_{\text{FS}},$$
$$F_s = \nu_\star [U_{11}(q) + U_{12}(q)] + F_t,$$
$$\tilde{F}_s = \nu_\star [U_{11}(q) - U_{12}(q)] + F_t. \tag{3.73}$$

Here $U_{11}(q) = 2\pi e^2/q\varepsilon$ and $U_{12}(q) = U_{11}(q)\exp(-qd)$. The quantities F_t and F_v are analogous to the standard Fermi liquid parameter in the triplet channel. They are determined by the screened interaction $U_{11/12}^{\text{scr}}(q, \omega)$ averaged over the Fermi surface. In the case of equal electron concentrations and mobilities in both quantum wells, we find

$$\langle U_{11/12}^{\text{scr}}(0) \rangle_{\text{FS}} = \int\limits_0^{2\pi} \frac{d\theta}{2\pi} U_{11/12}^{\text{scr}}(2k_F \sin(\theta/2), 0), \tag{3.74}$$

where k_F stands for the Fermi momentum in a quantum well. The interaction parameter F_s involves Coulomb interaction at a small transferred momentum. In the limit $q \to 0$, we obtain $F_s(q) \approx 2\varkappa/q \to \infty$ and $\tilde{F}_s = \varkappa d + F_t$. For $d = 0$, the double quantum well transforms into the single quantum well with $\tilde{F}_s = F_t = F_v$. In this case, the electron system is equivalent to the one in a Si-MOSFET. In the limit $d \to \infty$, intrawell electron–electron interaction vanishes and $\tilde{F}_s = F_s$ but $F_v = 0$. At finite value of d and $\Delta_s = 0$ the action becomes invariant under global $SU(2)$ spin rotations of electron operators in each well separately.

Dynamically screened inter- and intrawell interaction in RPA are given as (Zheng and MacDonald, 1993; Kamenev and Oreg, 1995; Flensberg et al., 1995)

$$U_{11}^{\text{scr}} = \frac{U_{11} + \Pi_2[U_{11}^2 - U_{12}^2]}{1 + [\Pi_1 + \Pi_2]U_{11} + \Pi_1\Pi_2[U_{11}^2 - U_{12}^2]},$$
$$U_{12}^{\text{scr}} = \frac{U_{12}}{1 + [\Pi_1 + \Pi_2]U_{11} + \Pi_1\Pi_2[U_{11}^2 - U_{12}^2]},$$
$$U_{22}^{\text{scr}} = \frac{U_{11} + \Pi_1[U_{11}^2 - U_{12}^2]}{U_{11} + \Pi_2[U_{11}^2 - U_{12}^2]} U_{11}^{\text{scr}}. \tag{3.75}$$

In the diffusive regime ($ql \ll 1$, $\omega\tau_{\text{tr}} \ll 1$) polarization operators become

$$\Pi_j(q, \omega) = v_\star \frac{D_j q^2}{D_j q^2 - i\omega}, \quad j = 1, 2, \tag{3.76}$$

where D_j denotes the diffusion coefficient in the j-th quantum well. We note that for $D_1 \neq D_2$ dynamically screened interwell interactions in the left and right quantum wells are different. In the case of equal electron concentrations and mobilities $D_1 = D_2 = D$, $U_{11}^{\mathrm{scr}} = U_{22}^{\mathrm{scr}}$ and

$$U_{11/12}^{\mathrm{scr}} = \frac{\varkappa(Dq^2 - i\omega)}{2v_\star q} \left\{ \frac{1 + e^{-qd}}{Dq\left[q + \varkappa(1 + e^{-qd})\right] - i\omega} \right.$$

$$\left. \pm \frac{1 - e^{-qd}}{Dq\left[q + \varkappa(1 - e^{-qd})\right] - i\omega} \right\}. \tag{3.77}$$

For $qd \gg 1$ electrons in the right quantum well do not affect interwell interaction in the left quantum well. In the opposite case, $qd \ll 1$, electrons in the right quantum well screen efficiently interaction between electrons in the left quantum well at $\varkappa d \lesssim 1$ only.

3.4.2.1 Estimates for interaction parameters

In the case of equal electron concentrations and mobilities the interaction parameters F_t and F_v can be estimated with the help of Eq. 3.75 as

$$F_t \pm F_v = -\int_0^{2\pi} \frac{d\theta}{4\pi} \frac{\varkappa(1 \pm e^{-2k_F d \sin\theta/2})}{2k_F \sin\frac{\theta}{2} + \varkappa(1 \pm e^{-2k_F d \sin\theta/2})}. \tag{3.78}$$

We note that RPA is justified provided $\varkappa/k_F \ll 1$. Equation 3.78 implies that F_t and F_v are negative and $|F_t| \geqslant |F_v|$. The interaction parameter $\tilde{F}_s$ changes sign from negative at small values of d to positive at large values of d. The dependence of d_c ($\tilde{F}_s(d_c) = 0$) on $\varkappa/k_F$ is shown in Fig. 3.8. We mention that $|\tilde{F}_s| \leqslant |F_t|$ at $d < d_c$.

In the case of the single quantum well the interaction parameter in the triplet channel (F_t^0) is given by Eq. 3.46. For quantum wells with equal electron concentrations and mobilities we find at

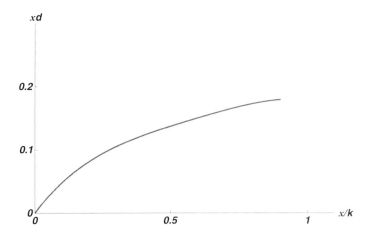

Figure 3.8 Dependence of $\varkappa d_c$ on $\varkappa / k_F$. Reprinted (figure) with permission from [Burmistrov, I. S., Gornyi, I. V., and Tikhonov, K. S. (2011). *Phys. Rev. B* **84**, p. 075338.] Copyright (2011) by the American Physical Society.

$k_F d \gg 1$:

$$F_t = F_t^0 + \frac{1}{8\pi k_F d}\mathcal{G}_1(\varkappa d), \qquad F_v = \frac{1}{8\pi k_F d}\mathcal{G}_2(\varkappa d),$$

$$\mathcal{G}_1(x) = \frac{3x\,e^x\,E_1(x)}{x+1} + \frac{2x\,e^{-2x/(x-1)}}{x^2-1}E_1\left(-\frac{2x}{x-1}\right),$$

$$\mathcal{G}_2(x) = \mathcal{G}_1(x) - \frac{4x\,e^x\,E_1(x)}{x+1}. \tag{3.79}$$

Here $E_1(x)$ denotes the integral exponent, $E_1(x) = \int_x^\infty dt\, \exp(-t)/t$.

3.4.3 One-Loop RG Equations

The matrix of interaction amplitudes Γ has the same structure as the matrix **F**. It is useful to introduce the following notations ($a = 1, 2, 3$ $b = 0, 1, 2, 3$):

$$\Gamma_{10} = z\tilde{\gamma}_s = -\frac{z\tilde{F}_s}{1+\tilde{F}_s}, \qquad \Gamma_{0a} = \Gamma_{1a} = z\gamma_t = -\frac{zF_t}{1+F_t},$$

$$\Gamma_{2b} = \Gamma_{3b} = z\gamma_v = -\frac{zF_v}{1+F_v}. \tag{3.80}$$

We remind that $\Gamma_{00} = -z$. Due to the presence of finite spin (Δ_s) and isospin (Δ_{SAS}) splittings the NLσM action Eq. 3.1 contains symmetry-breaking terms Eq. 3.5. In what follows we consider temperatures $T \gg \Delta_{s,SAS}$ or, equivalently, length scales $L \ll \sqrt{\sigma_{xx}/(16z\Delta_{s,SAS})}$ such that symmetry-breaking terms can be neglected.

Using Eqs. 3.80 and 3.24 we find the following one-loop RG equations for electrons in the symmetric double-quantum-well heterostructure (Burmistrov et al., 2011):

$$\frac{d\sigma_{xx}}{dy} = -\frac{2}{\pi}[2 + 1 + f(\tilde{\gamma}_s) + 6f(\gamma_t) + 8f(\gamma_v)], \tag{3.81}$$

$$\frac{d\tilde{\gamma}_s}{dy} = \frac{1 + \tilde{\gamma}_s}{\pi \sigma_{xx}}\Big[1 - 6\gamma_t - \tilde{\gamma}_s + 8\gamma_v + 2h(\tilde{\gamma}_s, \gamma_v)\Big], \tag{3.82}$$

$$\frac{d\gamma_t}{dy} = \frac{1 + \gamma_t}{\pi \sigma_{xx}}\Big[1 - \tilde{\gamma}_s + 2\gamma_t + h(\gamma_t, \gamma_v)\Big], \tag{3.83}$$

$$\frac{d\gamma_v}{dy} = \frac{1}{\pi \sigma_{xx}}\Big[(1 + \tilde{\gamma}_s)(1 - \gamma_v) + 2\gamma_v p(\gamma_t, \gamma_v)\Big], \tag{3.84}$$

$$\frac{d\ln z}{dy} = \frac{1}{\pi \sigma_{xx}}\Big[\tilde{\gamma}_s + 6\gamma_t + 8\gamma_v - 1\Big] \tag{3.85}$$

where $h(x, y) = 8y(x - y)/(1 + y)$ and $p(x, y) = 1 - 3x + 4y$. We emphasize that the right-hand side of Eqs. 3.82 and 3.83 is not polynomial in the intrawell interaction amplitude γ_v. In all other known examples, the right-hand side of one-loop RG equations for the interaction amplitudes are second-order polynomials (Finkelstein, 1990; Belitz and Kirkpatrick, 1994; Burmistrov and Chtchelkatchev, 2008; Punnoose, 2010a, 2010b). This is intimately related with invariance of the NLσM action $S_\sigma + S_F$ under transformations Eq. 3.6 of matrix Q. As follows from Eq. 3.7, the NLσM action $S_\sigma + S_F$ with interaction amplitudes Eq. 3.80 is invariant under rotations Eq. 3.6 with $\chi_{ab} = \chi \delta_{ac}\delta_{bd}$ where $c = 0, 1$, $d = 1, 2$, or 3 for $\gamma_t = -1$. For $\tilde{\gamma}_s = -1$ the NLσM action is invariant under rotations of Q matrix with $\chi_{ab} = \chi \delta_{a1}\delta_{b0}$. This guaranties that $\gamma_t = -1$ and $\tilde{\gamma}_s = -1$ are fixed under the action of RG transformations. Therefore, the right-hand side of RG equation for γ_t ($\tilde{\gamma}_s$) should vanish at $\gamma_t = -1$ ($\tilde{\gamma}_s = -1$). For $\gamma_v = -1$ the NLσM action $S_\sigma + S_F$ is not invariant under rotations Eq. 3.6. Therefore, the right-hand side of RG equations are not necessary finite at $\gamma_v = -1$.

RG equations (Eqs. 3.81–3.84) describe 4D $(\sigma_{xx}, \tilde{\gamma}_s, \gamma_t, \gamma_v)$ flow diagram. The 2D surface $\gamma_t = \gamma_v = \tilde{\gamma}_s$ is invariant under RG flow. It corresponds to the case of coinciding quantum wells $(d = 0)$. In this case, RG equations (Eqs. 3.81–3.85) are equivalent to RG equations (Eqs. 3.47–3.49). However, the surface $\gamma_t = \gamma_v = \tilde{\gamma}_s$ is unstable under perturbations in the initial values of interaction amplitudes due to finite value of d. There is the stable 2D surface $\gamma_v = 0$, $\tilde{\gamma}_s = -1$, which is invariant under RG flow. It corresponds to the case of separate quantum wells $(d = \infty)$.

In addition, the 2D surface $\gamma_t = \tilde{\gamma}_s = -1$ and the line $\tilde{\gamma}_s = -1$, $\gamma_v = -1/2$, $\gamma_t = -1/3$ are invariant under RG flow in the one-loop approximation. However, they are unreachable for the case of the double-quantum-well heterostructure since the following inequalities for the initial values of interaction amplitudes hold $\gamma_t(0) \geqslant \gamma_v(0) \geqslant 0$, and $\gamma_t(0) \geqslant \tilde{\gamma}_s(0)$. Using Eqs. 3.82–3.84, one can check that conditions $\gamma_t \geqslant \gamma_v \geqslant 0$ and $\gamma_t \geqslant \tilde{\gamma}_s$ remains true under the RG flow. Interaction amplitude γ_t increases always. Eventually, the RG flows toward $\gamma_v = 0$, $\tilde{\gamma}_s = -1$ and $\gamma_t = \infty$ as illustrated in Fig. 3.9.

As follows from RG equations (Eqs. 3.81–3.84) the temperature dependence of conductivity σ_{xx} is of metallic type. Depending on the sign of the quantity $2 + K_{ee}$ where $K_{ee} = 1 + f(\tilde{\gamma}_s(0)) + 6f(\gamma_t(0)) + 8f(\gamma_v(0))$, resistance either decreases monotonously under increase of y (for $2 + K_{ee} < 0$) or has the maximum (for $2 + K_{ee} > 0$).

3.4.4 Dephasing Time

The presence of electrons in the right quantum well changes the properties of electrons in the left quantum well. Using Eq. 3.31 in the case of symmetric double quantum well with equal electron concentrations and mobilities we find the dephasing rate for electrons in one quantum well for $1/\tau_\phi \gg \Delta_{s,SAS}$ (Burmistrov et al., 2011)

$$\frac{1}{\tau_\phi} = \frac{\mathcal{A}T}{2\sigma_{xx}} \ln T\tau_\phi, \quad \mathcal{A} = \frac{1}{2}\left[1 + \frac{\tilde{\gamma}_s^2}{2 + \tilde{\gamma}_s} + 6\frac{\gamma_t^2}{2 + \gamma_t} + 8\frac{\gamma_v^2}{2 + \gamma_v}\right].$$
$$(3.86)$$

Here factor $1/2$ appears in the quantity $\mathcal{A}$ since Eq. 3.86 is the dephasing rate for electrons in one quantum well. Interaction

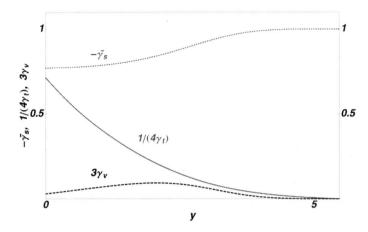

Figure 3.9 Dependence of γ_t, γ_v, and $\tilde{\gamma}_s$ on y for $\gamma_t(0) = 0.35$, $\gamma_v(0) = 0.01$, $\tilde{\gamma}_s(0) = -0.77$, and $\sigma_{xx}(0) = 6$.

amplitudes $\tilde{\gamma}_s$, γ_t, γ_v and conductivity σ_{xx} are taken at the length scale $L_T = \sqrt{\sigma_{xx}/zT}$.

It is instructive to compare result Eq. 3.86 with the result for the situation when one of the quantum wells is not filled by electrons. This corresponds to the limit $d \to \infty$. Using Eq. 3.86 with $\tilde{\gamma}_s = -1$, $\gamma_v = 0$, and $\gamma_t = \gamma_{t,0}$, we find (Narozhny et al., 2002)

$$\mathcal{A} \to \mathcal{A}_0 = \left[1 + \frac{3\gamma_{t,0}^2}{2 + \gamma_{t,0}}\right], \tag{3.87}$$

where initial value of $\gamma_{t,0}$ is equal to $\gamma_{t,0}(0) = -F_t^0/(1 + F_t^0)$.

3.4.5 Discussion and Comparison with Experiments

Recently, weak localization and interaction corrections to conductivity have been studied experimentally (Minkov et al., 2010, 2011) in an $Al_xGa_{1-x}As/GaAs/Al_xGa_{1-x}As$ double-quantum-well heterostructure. The detailed investigation of two heterostructures, 3243 and 3154, with different dopping level has been performed. Analysis of experimental data on magnetoresistance allows to extract the dephasing time and the quantity K_{ee}. The electron concentration in the right quantum well was controlled by a gate. In experiments of Minkov et al. (2010, 2011) the electron

Table 3.2 Parameters for samples of Minkov et al. (2010, 2011). For both samples, the inverse screening length and distance between quantum wells are equal to $\varkappa = 2.0 \times 10^6$ cm^{-1} and $d = 1.8 \times 10^{-6}$ cm, respectively

Sample	n, cm^{-2}	k_F, cm^{-1}	$k_F d$	$\varkappa/k_F$
#3154	4.5×10^{11}	1.7×10^6	3.06	1.18
#3243	7.5×10^{11}	2.2×10^6	3.95	0.91

concentration was high such that the total conductivity at high temperatures (about 4.2) was about $80\,e^2/h$. As a consequence, interesting physics described by RG equations (Eqs. 3.81–3.85) should develop in the heterostructures 3243 and 3154 at very low temperature only and, therefore, was not observed in experiments of Minkov et al. (2010, 2011). The unexpected result found by Minkov et al. (2010, 2011) is that the dephasing time (the coefficient $\mathcal{A}$ in Eq. 3.31) and interaction correction to the conductivity (the quantity K_{ee}) are almost independent of electron concentration in the right quantum well. The summary of experimental results of Minkov et al. (2010, 2011) is presented in Table 3.2. Theoretical estimates for the interaction parameters are summarized in Table 3.3. In the experiments parameter $\varkappa d$ can be estimated as $\varkappa d = 3.6$, interaction parameter F_v is small, $F_t \approx F_t^0$, and $\tilde{F}_s \approx \varkappa d$. Comparison of theoretical and experimental estimates for quantities K_{ee}, $\mathcal{A}$, $K_{ee,0} = 1 + 3f(\gamma_{t,0}(0))$ and $\mathcal{A}_0$ are shown in Table 3.4.

As one can see from the Table 3.4, theoretical estimates are in good quantitative agreement with the experimental findings. We note that since $K_{ee,0} > 0$ for $\varkappa/k_F \lesssim 1$, the interesting situation can be realized in the double-quantum-well heterostructure with

Table 3.3 Theoretical estimates for interaction amplitudes in samples 3154 and 3243 of Minkov et al. (2010, 2011)

Sample	$\tilde{\gamma}_s(0)$	$\gamma_t(0)$	$\gamma_v(0)$	$\gamma_{t,0}(0)$
#3154	−0.77	0.35	0.009	0.35
#3243	−0.77	0.30	0.007	0.30

Table 3.4 Comparison theoretical and experimental estimates for quantities K_{ee}, $\mathcal{A}$, $K_{ee,0}$, and $\mathcal{A}_0$

| | Theory | | Experiment | |
	#3154	#3243	#3154	#3243
K_{ee}	0.59	0.72	0.50 ± 0.05	0.57 ± 0.05
$K_{ee,0}$	0.52	0.59	0.53 ± 0.05	0.60 ± 0.05
$\mathcal{A}$	0.89	0.86		
$\mathcal{A}_0$	1.15	1.12		
$\mathcal{A}/\mathcal{A}_0$	0.77	0.77	1.00 ± 0.05	1.00 ± 0.05

$\varkappa d \lesssim 1$ in which in the case of equal electron concentration and mobilities one expects $K_{ee} < 0$.

The theory developed above is valid at temperatures $T \gg \Delta_{SAS}$, Δ_s, $1/\tau_{+-}$. In experiments of Minkov et al. (2010, 2011) the spin splitting (for magnetic fields in which magnetoresistance has been measured) and the symmetric-antisymmetric splitting can be estimated as $\Delta_s \lesssim 0.2$ K and $\Delta_{SAS} \lesssim 1$ K. The small asymmetry in impurity distribution along the z axis with respect to the middle plain between quantum wells yields elastic scattering between symmetric and antisymmetric states. The corresponding rate $(1/\tau_{+-})$ can be estimated from experimental data of Minkov et al. (2010, 2011) on magnetoresistance. As is well known (Altshuler and Aronov, 1985), in the absence of scattering between symmetric and antisymmetric states $(1/\tau_{+-} = 0)$, the presence of Δ_s and/or Δ_{SAS} does not influence weak localization correction. In the absence of a magnetic field the weak localization correction in both limiting cases, $\Delta_{SAS} \ll 1/\tau_{+-}$ and $\Delta_{SAS} \gg 1/\tau_{+-}$ can be written as (Burmistrov et al., 2011)

$$\delta\sigma_{xx}^{WL} = \frac{1}{\pi} \ln\left[\frac{\tau_{tr}^2}{\tau_\phi}\left(\frac{1}{\tau_\phi} + \frac{1}{\tau_{12}}\right)\right],$$

$$\frac{1}{\tau_{12}} \approx \min\left\{\Delta_{SAS}^2 \tau_{+-}, \tau_{+-}^{-1}\right\}. \tag{3.88}$$

Equation 3.88 interpolates between the result for two valleys at high temperatures $(1/\tau_\phi \gg 1/\tau_{12})$ and for a single valley at low temperatures $(1/\tau_\phi \ll 1/\tau_{12})$. For experiments of Minkov et al. (2010) the rate $1/\tau_{12}$ can be estimated as about 0.1 K. Using the estimate $\Delta_{SAS} \lesssim 1$ K, we find that $1/\tau_{+-} \approx 1/\tau_{12} \lesssim 0.1$ K. Therefore, the theory described above is applicable to the experiments of Minkov et al. (2010, 2011) at temperatures $T \gtrsim 1$ K. This is the range in which experiments of Minkov et al. (2010, 2011) have been performed.

3.5 Conclusions

Based on the NLσM approach and the RG treatment we study the effect of spin and isospin degrees of freedom on low-temperature transport in 2D strongly interacting disordered electron system. In this case we derive general RG equations in the one-loop approximation. We find that the standard case of equal interaction amplitudes for interaction among electrons with different spin and isospin projections is unstable. We demonstrate that such situation (with different interaction amplitudes) can be naturally realized in the 2D two-valley electron system in a Si-MOSFET and in the 2D electron system in a double-quantum-well heterostructure with common disorder. Using general one-loop RG equations we explain experimentally observed in a Si-MOSFET the variation of temperature dependence of resistivity from metallic to insulating with increase of the parallel magnetic field. For a Si-MOSFET we predict the temperature dependence of resistivity with two maximums. In a double-quantum-well heterostructure the general one-loop RG equations allow us to predict that in spite of common scatters electrons in each quantum well become independent at low temperatures.

A.1 Appendix

In this section we analyze the stability of the fixed point with $\Gamma_{ab} = \Gamma$ for $(ab) \neq (00)$ of the one-loop RG equation (Eq. 3.24). Let us write $\gamma_{ab} = \gamma + \eta_{ab}$ for $(ab) \neq (00)$ and $\gamma_{00} = \gamma_{00} + \eta_{00}$ where quantities η_{ab} are assumed to be small. Linearizing Eq. 3.24 we find

$$\frac{d\eta_{ab}}{dy} = \frac{1}{\pi\sigma_{xx}}\left[(17\gamma - \gamma_{00})\eta_{ab} - \sum_{cd}\eta_{cd}\left(\gamma\right.\right.$$

$$\left.\left. + \frac{1}{4}\,\mathrm{sp}[t_{ab}t_{cd}]^2\right)\right], \qquad (ab) \neq (00) \qquad (A.1)$$

and

$$\frac{d\eta_{00}}{dy} = -\frac{1}{\pi\sigma_{xx}}\left[(15\gamma + \gamma_{00})\eta_{00} + (1+\gamma_{00})\sum_{cd}\eta_{cd}\right]. \qquad (A.2)$$

It is convenient to introduce the following variables $\mu_{ab} = \sum_{cd\neq(00)}\mathrm{sp}[t_{ab}t_{cd}]^2\eta_{cd}/4$ and $\mu_{00} = \sum_{cd\neq(00)}\eta_{cd}$. We mention that μ_{00} is projection of the 16-dimensional vector $\{\eta_{ab}\}$ on the direction $\mathbf{e} = \{0, 1, \ldots, 1\}$. Then for $(ab) \neq (00)$ we obtain

$$\frac{d}{dy}\begin{pmatrix}\eta_{ab}\\\mu_{ab}\\\eta_{00}\\\mu_{00}\end{pmatrix} = \frac{1}{\pi\sigma_{xx}}\begin{pmatrix}\delta_1 & -1 & -\beta & -\gamma\\-16 & \delta_1 & \beta & \beta\\0 & 0 & -\delta_2 & -\beta_{00}\\0 & 0 & -15\beta & \delta_3\end{pmatrix}\begin{pmatrix}\eta_{ab}\\\mu_{ab}\\\eta_{00}\\\mu_{00}\end{pmatrix} \qquad (A.3)$$

where $\beta = 1+\gamma$, $\beta_{00} = 1+\gamma_{00}$, $\delta_1 = 17\gamma - \gamma_{00}$, $\delta_2 = 15\gamma + 2\gamma_{00} + 1$ and $\delta_3 = 2\gamma - \gamma_{00} + 1$. The behavior of quantities η_{ab} with $(ab) \neq (00)$ in the direction perpendicular to the plane based on vectors $\mathbf{e}$ and $\mathbf{e}_0 = \{1, 0, \ldots, 0\}$ is characterized by eigenvalues $\lambda_{\pm} = 17\gamma - \gamma_{00} \pm 4$. They are positive at large enough values of γ. This means that in Eq. A.3 small quantities η_{ab} satisfying $\eta_{00} = \mu_{00} = 0$ increase as the length scale y grows. Therefore the fixed point with $\Gamma_{ab} = \Gamma$ for $(ab) \neq (00)$ of the one-loop RG equations (Eq. 3.24) is unstable.

Acknowledgments

I am grateful to my coauthors N. M. Chtchelkatchev, I. V. Gornyi, D. A. Knyazev, O. E. Omel'yanovskii, V. M. Pudalov, and K. S. Tikhonov for fruitful collaboration on the subjects discussed in this review. I am indebted to A. V. Germanenko, D. A. Knyazev, A. A. Kuntsevich, G. M. Minkov, O. E. Omel'yanovskii, V. M. Pudalov, and A. A. Sherstobitov for detailed discussions of their experimental data prior to publication. And I thank A. S. Ioselevich, A. M. Finkelstein, A. D. Mirlin, P. M. Ostrovsky, A. M. M. Pruisken, M. A. Skvortsov, and A. G. Yashenkin for useful discussions and comments. The work was supported by Russian President grant no. MD-5620.2016.2, Russian President Scientific Schools grant NSh-10129.2016.2, and RFBR grant no. 15-32-20176.

References

Abrahams, E., and Ramakrishnan, T. V. (1980). *J. Non-Cryst. Solids* **35**, p. 15.

Abrahams, E., Anderson, P. W., and Ramakrishnan, T. V. (1979). *Phys. Rev. Lett.* **43**, p. 718.

Abrahams, E., Anderson, P.W., Licciardello, D. C., and Ramakrishnan, T. V. (1979). *Phys. Rev. Lett.* **42**, p. 673.

Abrahams, E., Kravchenko, S. V., and Sarachik, M. P. (2001). *Rev. Mod. Phys.* **73**, p. 251.

Altland, A., and Zirnbauer, M. R. (1997). *Phys. Rev. B* **55**, p. 1142.

Altshuler, B. L., and Aronov, A. G. (1981). *JETP Lett.* **33**, p. 499.

Altshuler, B. L., Aronov, A. G., Larkin, A. I., and Khmel'nitskii, D. E. (1981). *JETP* **54**, p. 411.

Altshuler, B. L., and Aronov, A. G. (1979a). *JETP* **50**, p. 968.

Altshuler, B. L., and Aronov, A. G. (1979b). *JETP Lett.* **30**, p. 482.

Altshuler, B. L., and Aronov, A. G. (1985). *Electron-Electron Interactions in Disordered Conductors*, eds. A. J. Efros and M. Pollack (Elsevier Science, North-Holland).

Altshuler, B. L., Aronov, A. G., and Khmelnitsky, D. E. (1982). *J. Phys. C* **15**, p. 7367.

Altshuler, B. L., Aronov, A. G., and Lee, P. A. (1980). *Phys. Rev. Lett.* **44**, p. 1288.

Altshuler, B. L., Khmel'nitzkii, D. E., Larkin, A. I., and Lee, P. A. (1982). *Phys. Rev. B* **22**, p. 5142.

Amit, D. J. (1984). *Field Theory, Renormalization Group, and Critical Phenomena* (World Scientific).

Anderson, P.W. (1958). *Phys. Rev.* **109**, p. 1492.

Ando, T., Fowler, A. B., and Stern, F. (1982). *Rev. Mod. Phys.* **54**, p. 437.

Anissimova, S., Kravchenko, S. V., Punnoose, A., Finkel'stein, A. M., and Klapwijk, T. M. (2007). *Nat. Phys.* **3**, p. 707.

Baranov, M. A., Burmistrov, I. S., and Pruisken, A. M. M. (2002). *Phys. Rev. B* **66**, p. 075317.

Baranov, M. A., Pruisken, A. M. M., and Škorić, B. (1999). *Phys. Rev. B* **60**, p. 16821.

Basko, D. M., Aleiner, I. L., and Altshuler, B. L. (2006). *Ann. Phys. (N.Y.)* **321**, p. 1126.

Belitz, D., and Kirkpatrick, T. R. (1994). *Rev. Mod. Phys.* **66**, p. 261.

Berezinskii, V. L. (1974). *JETP* **38**, p. 620.

Blanter, Ya. M. (1996). *Phys. Rev. B* **54**, p. 12807.

Boebinger, G. S., Jiang, H.W., Pfeiffer, L. N., andWest, K.W. (1990). *Phys. Rev. Lett.* **64**, p. 1793.

Br'ezin, E., Hikami, S., and Zinn-Justin, J. (1980). *Nucl. Phys. B* **165**, p. 528.

Brener, S., Iordanski, S. V., and Kashuba, A. (2003). *Phys. Rev. B* **67**, p. 125309.

Burmistrov, I. S., and Chtchelkatchev, N. M. (2007). *JETP Lett.* **84**, p. 656.

Burmistrov, I. S., and Chtchelkatchev, N. M. (2008). *Phys. Rev. B* **77**, p. 195319.

Burmistrov, I. S., Gornyi, I. V., and Tikhonov, K. S. (2011). *Phys. Rev. B* **84**, p. 075338.

Castellani, C., and Di Castro, C. (1986). *Phys. Rev. B* **34**, p. 5935.

Castellani, C., Di Castro, C., and Lee, P. A. (1998). *Phys. Rev. B* **57**, p. 9381.

Castellani, C., Di Castro, C., Lee, P. A., and Ma, M. (1984a). *Phys. Rev. B* **30**, p. 527.

Castellani, C., Di Castro, C., Lee, P. A., Ma, M., Sorella, S., and Tabet, E. (1984b). *Phys. Rev. B* **30**, p. 1596.

Dell'Anna, L. (2006). *Nucl. Phys. B* **758**, p. 255.

Dobrosavljević, V. (2010). *Int. J. Mod. Phys. B* **24**(12&13).

Dyson, F. J. (1962a). *J. Math. Phys.* **3**, p. 1199.

Dyson, F. J. (1962b). *J. Math. Phys.* **3**, p. 140.

Efetov, K. B. (1982). *JETP* **55**, p. 514.

Efetov, K. B. (1983). *Adv. Phys.* **32**, p. 53; (1997). *Supersymmetry in Disorder and Chaos* (Cambridge University Press).

Efetov, K. B., Larkin, A. I., and Kheml'nitskii, D. E. (1980). *JETP* **52**, p. 568.

Eisenstein, J. P., and MacDonald, A. H. (2004). *Nature* **432**, p. 691.

Finkelstein, A. M. (1983a). *JETP* **57**, p. 97.

Finkelstein, A. M. (1983b). *JETP Lett.* **37**, p. 517.

Finkelstein, A. M. (1984a). *JETP* **59**, p. 212.

Finkelstein, A. M. (1984b). *JETP Lett.* **40**, p. 796.

Finkelstein, A. M. (1984c). *Z. Phys. B* **56**, p. 189.

Finkelstein, A. M. (1990). *Electron Liquid in Disordered Conductors*, Vol. 14 of *Soviet Scientific Reviews*, ed. I. M. Khalatnikov (Harwood Academic, London).

Fleishman, L., and Anderson, P.W. (1980). *Phys. Rev. B* **21**, p. 2366.

Flensberg, K., Yu-Kuang-Hu, B., Jauho, A.-P., and Kinaret, J. M. (1995). *Phys. Rev. B* **52**, p. 14761.

Fröhlich, J.,Martinelli, F., Scoppola, E., and Spencer, T. (1985). *Commun. Math. Phys.* **101**, p. 21.

Gantmakher, V. F., and Dolgopolov, V. T. (2008). *Phys. Usp.* **51**, p. 3.

Gorkov, L. P., Larkin, A. I., and Khmel'nitskii, D. E. (1979). *JETP Lett.* **30**, p. 228.

Gornyi, I. V., Mirlin, A. D., and Polyakov, D. G. (2005). *Phys. Rev. Lett.* **95**, p. 206603.

Gramila, T. J., Eisenstein, J. P., MacDonald, A. H., Pfeiffer, L. N., and West, K.W. (1991). *Phys. Rev. Lett.* **66**, p. 1216.

Gunawan, O., Gokmen, T., Vakili, K., Padmanabhan, M., De Poortere, E. P., and Shayegan, M. (2007). *Nat. Phys.* **3**, p. 388.

Gunawan, O., Shkolnikov, Y. P., Vakili, K., Gokmen, T., De Poortere, E. P., and Shayegan, M. (2006). *Phys. Rev. Lett.* **97**, p. 186404.

Heinzner, P., Huckleberry, A., and Zirnbauer, M. R. (2005). *Commun. Math. Phys.* **257**, p. 725.

Jüngling, K., and Oppermann, R. (1980). *Z. Phys. B* **38**, p. 93.

Jendrzejewski, F., Bernard, A., Müller, K., Cheinet, P., Josse, V., Piraud, M., Pezzé, L., Sanchez-Palencia, L., Aspect, A., and Bouyer, P. (2012). *Nat. Phys.* **8**, p. 398.

Kamenev, A., and Andreev, A. (1999). *Phys. Rev. B* **60**, p. 2218.

Kamenev, A., and Levchenko, A. (2009). *Adv. Phys.* **58**, p. 197.

Kamenev, A., and Oreg, Y. (1995). *Phys. Rev. B* **52**, p. 7516.

Kellog, M., Eisenstein, J. P., Pfeiffer, L. N., and West, K. W. (2004). *Phys. Rev. Lett.* **93**, p. 036801.

Khrapai, V. S., Deviatov, E. V., Shashkin, A. A., Dolgopolov, V. T., Hastreiter, F., Wixforth, A., Campman, K. L., and Gossard, A. C. (2000). *Phys. Rev. Lett.* **84**, p. 725.

Kitaev, A. Yu. (2009). *AIP Conf. Proc.* **1134**, p. 22.

Klimov, N. N., Knyazev, D. A., Omelyanovskii, O. E., Pudalov, V.M., Kojima, H., and Gershenson, M. E. (2008). *Phys. Rev. B* **78**, p. 195308.

Knyazev, D. A., Omel'yanovskii, O. E., Pudalov, V. M., and Burmistrov, I. S. (2006). *JETP Lett.* **84**, p. 662.

Knyazev, D. A., Omel'yanovskii, O. E., Pudalov, V. M., and Burmistrov, I. S. (2008). *Phys. Rev. Lett.* **100**, p. 46405.

Kravchenko, S. V., and Sarachik, M. P. (2004). *Rep. Prog. Phys* **67**, p. 1.

Kravchenko, S. V., Kravchenko, G. V., Furneaux, J. E., Pudalov, V. M., and D'Iorio, M. (1994). *Phys. Rev. B* **50**, p. 8039.

Kravchenko, S. V., Mason, W. E., Bowker, G. E., Furneaux, J. E., Pudalov, V. M., and D'Iorio, M. (1995). *Phys. Rev. B* **51**, p. 7038.

Kuntsevich, A. Yu., Klimov, N. N., Tarasenko, S. A., Averkiev, N. S., Pudalov, V. M., Kojima, H., and Gershenson, M. E. (2007). *Phys. Rev. B* **75**, p. 195330.

Landau, L. D., and Lifshitz, E. M. (1991). *Quantum Mechanics, Course of Theoretical Physics*, Vol. 3 (Pergamon).

Lee, P. A., and Ramakrishnan, T. V. (1985). *Rev. Mod. Phys.* **57**, p. 287.

Levine, H., Libby, S. B., and Pruisken, A. M. M. (1983). *Phys. Rev. Lett.* **51**, p. 1915.

Lian Zheng and MacDonald, A. H. (1993). *Phys. Rev. B* **48**, p. 8203.

Lilly, M. P., Eisenstein, J. P., Pfeiffer, L. N., and West, K. W. (1998). *Phys. Rev. Lett.* **80**, p. 1714.

Lopez, M., Clément, J.-F., Szriftgiser, P., Garreau, J. C., and Delande, D. (2012). *Phys. Rev. Lett.* **108**, p. 095701.

McKane, A. J., and Stone, M. (1981). *Ann. Phys. (N.Y.)* **131**, p. 36.

McMillan,W. L. (1981). *Phys. Rev. B* **24**, p. 2739.

Minkov, G. M., Germanenko, A. V., Rut, O. E., Khrykin, O. I., Shashkin, V. I., and Daniltsev, V. M. (2000). *Nanotechnology* **11**, p. 406.

Minkov, G. M., Germanenko, A. V., Rut, O. E., Sherstobitov, A. A., Bakarov, K., and Dmitriev, D. V. (2010). *Phys. Rev. B* **82**, p. 165325.

Minkov, G. M., Germanenko, A. V., Rut, O. E., Sherstobitov, A. A., Bakarov, K., and Dmitriev, D. V. (2011). *Phys. Rev. B* **84**, p. 075337.

Mirlin, A. D. (2000). *Phys. Rep.* **326**, p. 259.

Mirlin, A. D., and Evers, F. (2008). *Rev. Mod. Phys.* **80**, p. 1355.

Mott, N. F., and Twose,W. D. (1961). *Adv. Phys.* **10**, p. 107.

Narozhny, B. N., Zala, G., and Aleiner, I. L. (2002). *Phys. Rev. B* **65**, p. 180202.

Nestoklon,M. O., Golub, L. E., and Ivchenko, E. I. (2006). *Phys. Rev. B* **73**, p. 235334.

Pagnossin, I. R., Meikap, A. K., Lamas, T. E., Gusev, G. M., and Portal, J. C. (2008). *Phys. Rev. B* **78**, p. 115311.

Pillarisetty, R., Hwayong Noh, Tsui, D. C., De Poortere, E. P., Tutuc, E., and Shayegan, M. (2002). *Phys. Rev. Lett.* **89**, p. 016805.

Polyakov, A. M. (1975). *Phys. Lett. B* **59**, p. 79.

Pruisken, A. M. M. (1984). *Nucl. Phys. B* **235**, p. 277.

Pruisken, A. M. M. (1987). *The Quantum Hall Effect*, p. 117, eds. R. E. Prange and S. M. Girvin (Springer).

Pruisken, A. M. M., Baranov, M. A., and Burmistrov, I. S. (2005). *JETP Lett.* **82**, p. 150.

Pruisken, A.M.M., Baranov, M. A., and Škorić, B. (1999). *Phys. Rev. B* **60**, p. 16807.

Pudalov, V. M., Gershenson, M. E., and Kojima, H. (2004). On the electron-electron interactions in two dimensions, in *Fundamental Problems of Mesoscopic Physics. Interaction and Decoherence*, pp. 309–327, *Nato Science Series*, eds. I. V. Lerner, B. L. Altshuler, and Y. Gefen (Kluwer).

Pudalov, V. M., Gershenson, M. E., Kojima, H., Brunthaler, G., Prinz, A., and Bauer, G. (2003). *Phys. Rev. Lett.* **91**, p. 126403.

Punnoose, A. (2010a). *Phys. Rev. B* **81**, p. 035306.

Punnoose, A. (2010b). *Phys. Rev. B* **82**, p. 115310.

Punnoose, A., and Finkelstein, A. M. (2001). *Phys. Rev. Lett.* **88**, p. 016802.

Punnoose, A., and Finkelstein, A. M. (2005). *Science* **310**, p. 289.

Punnoose, A., Finkel'stein, A. M., Mokashi, A., and Kravchenko, S. V. (2010). *Phys. Rev. B* **82**, p. 201308(R).

Renard, V. T., Duchemin, I., Niida, Y., Fujiwara, A., Hirayama, Y., and Takashina, K. (2013). *Sci. Rep.* **3**, p. 2011.

Sawada, A., Ezawa, Z. F., Ohno, H., Horikoshi, Y., Ohno, Y., Kishimoto, S., Matsukura, F., Yasumoto, M., and Urayama, A. (1999). *Phys. Rev. Lett.* **80**, p. 4534.

Schäfer, L., and Wegner, F. (1980). *Z. Phys. B* **38**, p. 113.

Schmidt, A. (1974). *Z. Phys. B* **271**, p. 251.

Schnyder, A. P., Ryu, S., Furusaki, A., and Ludwig, A. W. W. (2008). *Phys. Rev. B* **78**, p. 195125.

Schnyder, A. P., Ryu, S., Furusaki, A., and Ludwig, A. W. W. (2009). *AIP Conf. Proc.* **1134**, p. 10.

Shangina, E. L., and Dolgopolov, V. T. (2003). *Phys. Usp.* **46**, p. 777.

Shashkin, A. A. (2005). *Phys. Usp.* **48**, p. 129.

Shayegan, M., De Poortere, E. P., Gunawan, O., Shkolnikov, Y. P., Tutuc, E., and Vakili, K. (2006). *Phys. Status Solidi (b)* **243**, p. 3629.

Simonian, D., Kravchenko, S. V., Sarachik, M. P., and Pudalov, V. M. (1997). *Phys. Rev. Lett.* **79**, p. 2304.

Sivan, U., Solomon, P. M., and Shtrikman, H. (1992). *Phys. Rev. Lett.* **68**, p. 1196.

Takashina, K., Niida, Y., Renard, V. T., Fujiwara, A., Fujisawa, T., Muraki, K., and Hirayama, Y. (2011). *Phys. Rev. Lett.* **106**, p. 196403.

Thouless, D. J. (1974). *Phys. Rep.* **13**, p. 93.

Thouless, D. J. (1977). *Phys. Rev. Lett.* **39**, p. 1167.

Tutuc, E., Shayegan, M., and Huse, D. A. (2004). *Phys. Rev. Lett.* **93**, p. 036802.

Vitkalov, S. A., James, K., Narozhny, B. N., Sarachik, M. P., and Klapwijk, T. M. (2003). *Phys. Rev. B* **67**, p. 113310.

Wegner, F. (1976). *Z. Phys. B* **25**, p. 327.

Wegner, F. (1979). *Z. Phys. B* **35**, p. 207.

Wigner, E. P. (1951). *Ann. Math.* **53**, p. 36.

Yoon, J., Li, C. C., Shahar, D., Tsui, D. C., and Shayegan, M. (2000). *Phys. Rev. Lett.* **84**, p. 4421.

Zala, G., Narozhny, B. N., and Aleiner, I. L. (2001a). *Phys. Rev. B* **64**, p. 214204.

Zala, G., Narozhny, B. N., and Aleiner, I. L. (2001b). *Phys. Rev. B* **65**, p. 020201.

Zinn-Justin, J. (1989). *Quantum Field Theory and Critical Phenomena* (University Press).

Zirnbauer, M. R. (1996). *J. Math. Phys.* **37**, p. 4986.

Chapter 4

Electron Transport Near the 2D Mott Transition

Tetsuya Furukawa and Kazushi Kanoda

Department of Applied Physics, University of Tokyo, Bunkyo, Tokyo 113-0035, Japan
kanoda@ap.t.u-tokyo.ac.jp

4.1 Mott Transition

Mutual interactions between electrons afford various kinds of exotic electronic phases such as high-T_c superconductivity, unconventional magnetism, colossal magnetoresistance, and so on (Imada et al., 1998). The interaction is represented by the Coulomb repulsive energy for two electrons accommodated by a site, which is called the on-site Coulomb energy U. On the other hand, electrons tend to propagate as Bloch waves on a periodic lattice owing to the wave-like nature by gaining quantum-mechanical kinetic energy characterized by the electronic bandwidth W. In the simple case of a half-filled band, the competition between the Coulomb energy and kinetic energy causes a metal–insulator transition; that is, the Mott transition. For $W \gg U$, electrons behave itinerant with the double occupation allowed, whereas for $W \ll U$, electrons keep staying

Strongly Correlated Electrons in Two Dimensions
Edited by Sergey Kravchenko
Copyright © 2017 Pan Stanford Publishing Pte. Ltd.
ISBN 978-981-4745-37-6 (Hardcover), 978-981-4745-38-3 (eBook)
www.panstanford.com

at each site to avoid the double occupation. A dramatic metal–insulator transition (the Mott transition) is expected to occur when $W \approx U$ as the interaction-induced imbalance between the wave and particle natures. This phenomenon is one of the most prominent manifestations of the electron correlation in condensed matters and has long been studied since Mott's argument on the insulating state of NiO_2 (Mott, 1990). The Mott transition is recognized as a key concept for understanding diverse interaction-induced emergent phenomena.

Usually, phase transition is almost always accompanied by symmetry breaking, of which the degree is characterized by an order parameter. However, there is no symmetry breaking in the Mott transition, which occurs in the charge degrees of freedom in principle, although in many cases, some orders occur in other degrees of freedom such as spins, lattices, and orbitals at low temperatures. Because any phase transition without symmetry breaking should be a crossover at high temperatures, the temperature (T)-interaction (represented by U/W) phase diagram of the Mott transition is expected to look like Fig. 4.1 as in the gas–liquid transition. In this chapter, we look into the critical behavior of the Mott transition in

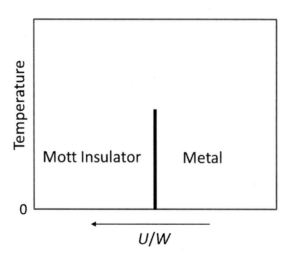

Figure 4.1 Conceptual phase diagram of Mott transition by analogy to the gas–liquid transition.

organic conductors, which are superior model materials for the Mott physics.

4.2 Theoretical Investigations of the Mott Transition

4.2.1 Dynamical Mean Field Theory

To treat the physics of the Mott transition theoretically, the Hubbard model (Hubbard, 1964), the minimal model of correlated electrons on a lattice, is useful. The model has two competing terms: a kinetic energy term and an on-site Coulomb energy term. The former is characterized by transfer integrals between different sites. In many cases, only the transfer integrals between nearest-neighbor sites t and those between next-nearest-neighbor sites t' are considered. The latter term is characterized by on-site Coulomb repulsive energy U. Thus, the Hubbard model is written as,

$$H = t \sum_{i,j(n.n.)} c_{i,\sigma}^+ c_{j,\sigma} + t' \sum_{i,j(n.n.n.)} c_{i,\sigma}^+ c_{j,\sigma} + U \sum_i c_{i,\uparrow}^+ c_{i,\uparrow} c_{i,\downarrow}^+ c_{i,\downarrow}$$

where $c_{i,\sigma}^+$ and $c_{i,\sigma}$ are creation and annihilation operators of electrons with spin σ at the lattice site labeled by i, and (n.n.) and (n.n.n.) stand for pairs of nearest neighbors and next-nearest neighbors, respectively. The Hubbard model in a half-filled system can derive a metallic (Fermi liquid) phase for $t, t' \gg U$ and a Mott insulating phase for $U \gg t, t'$. The Mott transition occurs between the two phases when $t/U \approx W/U$ is varied.

Theoretical treatment of the Mott transition is not straight-forward, even if one works with the Hubbard model as the minimal model of the Mott transition. This is because the conventional perturbative approaches completely collapse near the Mott transition, in which W and U are comparable. The first theoretical breakthrough was achieved by Brinkman and Rice in the early 1970s (Brinkman and Rice, 1970). They used a variational method and proposed the scenario that a quasiparticle's mass in the metallic phase diverges toward the Mott transition. Although their study can describe only the ground state of metallic phase, this Brinkman–Rice scenario was quite convincing. Indeed, more developed and innovative theories, such as dynamical mean field theory (DMFT) (Zhang et al., 1993; Georges et al., 1996), have partially supported the Brinkman–Rice scenario, and DMFT has provided a clear description of the Mott transition in the last two decades. In DMFT the problem in a system of correlated electrons on a lattice is mapped to the problem of an impurity embedded in an effective medium. This mapping allows us to treat a local interaction of electrons, namely on-site Coulomb repulsion U, with a nonperturbative manner and yields rigorous results for an infinite-dimensional system by taking into account all local quantum fluctuations. One of the great achievements made by DMFT is the full description of the electronic density of states across the Mott transition, as shown in Fig. 4.2: the width of a quasiparticle coherence peak in the metallic phase get narrower on approaching the Mott transition and vanishes on crossing it, as Brinkman and Rice predicted. In addition, the separation of the upper and lower incoherent parts of the density of states in a metallic phase results in the formation of the upper and lower Hubbard bands with a Mott–Hubbard gap at the transition.

The DMFT also provided a temperature correlation phase diagram, which showed that the first-order Mott transition terminated at the finite-temperature critical end point (Georges et al., 1996; Terletska et al., 2011). At temperatures lower than the critical temperature T_c, the Mott transition is a discontinuous phase transition, whereas the transition crossing T_c is continuous. In general, a continuous phase transition is accompanied with a critical phenomenon; thus, the phenomenon is expected around the critical end point in the case of the Mott transition (Kotliar et al., 2000). Noticeably, the critical phenomenon was experimentally observed around the critical end point of the Mott transitions in Cr-doped V_2O_3 (Limelette, 2003b) and organic charge transfer salts κ-$(ET)_2X$ (Kagawa et al., 2005, 2009; Bartosch et al., 2010). We see the experimental results in the following Section 4.5. Note that the system can move from the metallic phase to the Mott insulating phase without thermodynamic singularities, making a detour to avoid the first-order line and the critical point. This character clearly indicates that the Mott transition occurs without symmetry

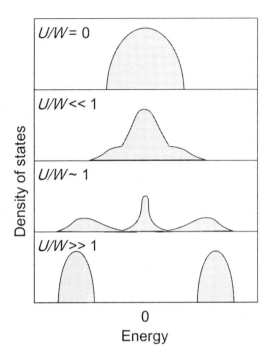

Figure 4.2 Density of states at $T = 0$ calculated by DMFT for several values of U/W. The upper three spectra are for the metal, while the bottom one is for the Mott insulator.

breaking. Moreover, recent studies based on cluster DMFT (CDMFT) (Parcollet et al., 2004), which is an extended version of DMFT taking into account the spatial correlation neglected in the original DMFT, have a phase diagram reminiscent of that of a real system, as seen in later sections (Park et al., 2008; Liebsch et al., 2009).

4.2.2 Quantum Criticality of the Mott Transition

Classical phase transition and concomitant critical phenomenon around a critical point between high- and low-temperature phases is driven by thermal fluctuations, where the former phase has larger entropy than the latter. In contrast, quantum phase transition and accompanied quantum criticality are driven by quantum fluctuations between two competing ground states (Sachdev, 2011; Sondhi et al., 1997). For the case of the quantum-critical phenomena

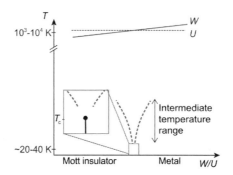

Figure 4.3 Schematic phase diagram involving quantum criticality of the Mott transition. If bandwidth W and on-site Coulomb repulsion energy U are sufficiently larger than the critical temperature T_c, there is a wide intermediate-temperature range ($T_c < T \ll W, U$) in which a quantum criticality of the Mott transition can be expected. Temperature scales displayed are for the case of κ-$(ET)_2 X$.

of the Mott transition, competing energies are on-site Coulomb repulsion energy U and bandwidth W. Quantum-critical fluctuations due to the competition can be enhanced at temperatures well below these energy scales. Hence, even if the critical temperature of the Mott transition, T_c, is finite, unlike the genuine quantum phase transition, in which $T_c = 0$ K, in case T_c is orders of magnitude lower than W and U, there is a wide temperature region of $T_c < T \ll U$, W, where the system is subjected to quantum fluctuations (Fig. 4.3) and exhibits a quantum criticality of the Mott transition without the effect of thermal critical fluctuations around the finite-temperature critical point.

In line with the above idea, Terletska et al. performed the scaling analysis of the resistivities $\rho(T, U)$ calculated by DMFT (Terletska et al., 2011; Vučičević et al., 2013) for various T and U. They determined the quantum Widom line of the Mott transition $U_c(T)$, where the free energy is considered to be most unstable against U at temperatures above T_c, and defined the distance from the quantum Widom line as $\delta U = U - U_c(T)$. Then, they showed that a set of calculated resistivity curves satisfies the scaling relation of $\rho(T, \delta U) = \rho_c(T) f_\pm[T/T_0(\delta U)]$, where $\rho_c(T) = \rho(T, \delta U = 0)$ is the critical resistivity, $f_\pm$ are scaling functions for the insulating branch $(+, \delta U > 0)$ and the metallic branch $(-, \delta U < 0)$, $T_0(\delta U) =$

$|\delta U|^{z\nu}$, and z and ν are the critical exponents. The values of $z\nu$ were 0.56 ± 0.01 and 0.57 ± 0.01 for the metallic and the insulating branch, respectively (Terletska et al., 2011). The fulfillment of this scaling relation is a property inherent to a quantum criticality: physical quantities in quantum-critical region are determined by only the ratio of absolute temperature T to the characteristic energy scale $T_0(\delta U)$, which vanishes on approaching $\delta U = 0$. Surprisingly, the proposed quantum criticality of the Mott transition at the intermediate temperatures ($T_c < T \ll U$, W) has been observed in organic materials (Furukawa et al., 2015). This topic is explained in Section 4.6.

4.3 Organic Materials: Model Systems of the Mott Physics

To investigate the Mott transition experimentally, particularly to look into the criticality of the Mott transition, one needs real materials with half-filled bands of controllable widths in the vicinity of the Mott transition. There are several systems that serve the purpose, such as V_2O_3 (McWhan et al., 1973), $NiS_{2-x}Se_x$ (Yao et al., 1996), and organic materials (Kanoda, 1997a, 1997b, 2006; Kanoda and Kato, 2011). Among them, organic materials are advantageous in the following respects: (i) Irrespective of the complex structures, the electronic bands are simple because the tight-binding model of molecular orbitals works well (e.g., Mori et al., 1984); (ii) a variety of molecular arrangements allows experimental studies with the lattice geometry as a parameter; (iii) the van der Waals nature of molecular packing enables one to vary the transfer integrals, namely bandwidth W, in a wide range by chemical substitution or physical pressure. Thanks to these features, diverse electronic phases emerge under geometrical frustration and electron correlation varied by hydrostatic or uniaxial pressures, and the physics behind them has been argued on the basis of the band structures in a systematic way (Kino and Fukuyama, 1995, 1996).

Quasi-2D organic charge transfer salts, $\kappa\text{-}(ET)_2X$ and $Y[Pd(dmit)_2]_2$, are half-filled band systems with anisotropic tri-

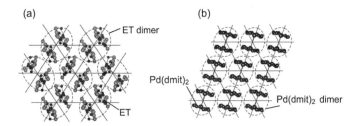

Figure 4.4 Molecular arrangements in the conducting layer in (a) κ-$(ET)_2X$ and (b) $Y[Pd(dmit)_2]_2$.

angular lattices (Kanoda and Kato, 2011). ET and dmit represent bis(ethylenedithio) tetrathiafulvalene and 1,3-dithiole-2-thione-4,5-dithiolate, respectively. Both materials have layered structures, where $(ET)^{+0.5}$ and $[Pd(dmit)_2]^{-0.5}$ form conducting layers and monovalent anion X^{-1} and cation Y^{+1} with closed shells form insulating layers in κ-$(ET)_2X$ and $Y[Pd(dmit)_2]_2$, respectively. The Greek character κ represents the specific packing pattern of ET molecules, as shown in Fig. 4.4a. In κ-$(ET)_2X$, the conducting layers are constructed by the quasi-triangular arrangement of ET dimers, each of which accommodates a hole in an antibonding dimer orbitals constructed by the highest occupied molecular orbital (HOMO) of ET. Thus, κ-$(ET)_2X$ are half-filled band systems with anisotropic triangular lattices. The band structure is schematically shown in Fig. 4.5a. Because a unit cell contains two dimers, the Fermi surface is folded at the boundary of the first Brillouin zone but is essentially

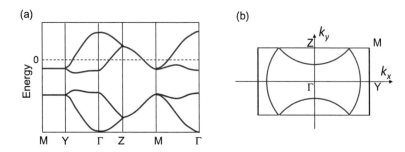

Figure 4.5 (a) Band structure and (b) Fermi surfaces in the first Brilluoin zone for κ-$(ET)_2X$ (Mori et al., 1984).

a simple cylinder, as shown in Fig. 4.5b. In $Y[Pd(dmit)_2]_2$, the packing patterns of $Pd(dmit)_2$ are different from the κ-type (see Fig. 4.4b); this pattern is called β'-type (Kanoda and Kato, 2011; Kato and Hengbo, 2012), and the quasi-degenerate nature of HOMO and LUMO (lowest unoccupied molecular orbital) in $Pd(dmit)_2$ makes the orbital hybridization in a dimer, $[Pd(dmit)_2]_2$, more complicated than in $(ET)_2$. Nevertheless, a half-filled band and a cylindrical Fermi surface are hosted by this material as well because of the dimeric and triangular natures in the molecular arrangements (Kato, 2014). The substitution of X and Y or the application of an external pressure varies the effective electronic correlation U/t and the triangularity t'/t, where t and t' are the transfer integrals of the nearest and the next-nearest-neighbor dimers, respectively.

4.4 Temperature–Pressure Phase Diagram

High compressibility due to the van der Walls nature of molecular packing in κ-$(ET)_2X$ and $Y[Pd(dmit)_2]_2$ allows us to control bandwidth W of a system by applying pressure; indeed, these organic systems exhibit the pressure-induced bandwidth-controlled Mott transition. In particular, the Mott transition in κ-$(ET)_2Cu[N(CN)_2]Cl$ and κ-$(ET)_2Cu_2(CN)_3$, which are abbreviated by κ-Cl and κ-$Cu_2(CN)_3$, respectively, have been studied intensively so far. Figure 4.6 shows the temperature–pressure phase diagrams for κ-Cl (Lefebvre et al., 2000; Limelette et al., 2003a; Kagawa et al., 2004) and κ-$Cu_2(CN)_3$ (Kurosaki et al., 2005; Kobashi, 2007). For both phase diagrams, the Mott insulating phase and the metallic phase are separated by the first-order Mott transition line, which ends at a finite-temperature critical end point. Above the critical point, the Mott transition becomes a metal–insulator crossover. The critical temperatures are 40 K for κ-Cl and approximately 20 K for κ-$Cu_2(CN)_3$. In the vicinity of the critical point of κ-$(ET)_2X$, many physical quantities, such as conductivity (Kagawa et al., 2005), sound velocity (Fournier et al., 2003), thermal expansion (Bartosch et al., 2010; De Souza et al., 2007), and nuclear magnetic resonance (Kagawa et al., 2009), show anomalies. At low temperatures in the insulating phase, the antiferromagnetic long-range order with

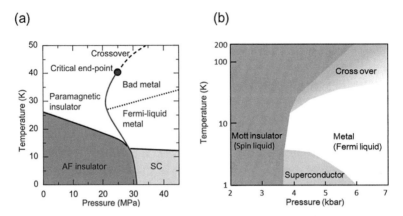

Figure 4.6 Temperature–pressure phase diagrams of κ-Cl (a) (Kagawa et al., 2005) and κ-Cu$_2$(CN)$_3$ (b) (Kurosaki et al., 2005). In the left panel, AF and SC denote the antiferromagnetic order and the superconductor, respectively, and the bold line is the first-order Mott transition line. In the right panel, below approximately 20 K, the phase boundary between the Mott insulator phase and the metal and superconductor phases is the first-order transition line. Modified (figure) with permission from [Kurosaki, Y., Shimizu, Y., Miyagawa, K., Kanoda, K., and Saito, G. (2005). *Phys. Rev. Lett.* **95**, p. 177001.] Copyright (2005) by the American Physical Society.

an ordered moment of 0.45 μ_B (μ_B: Bohr magneton) appears in κ-Cl (Miyagawa et al., 1995). On the other hand, for κ-Cu$_2$(CN)$_3$, magnetic order is absent, and quantum spin liquid phase emerges owing to the strong geometrical frustration coming from the nearly isotropic triangular lattice (Shimizu et al., 2003; Shimizu et al., 2006) with $t'/t \approx 0.8$–1.0 (Kandpal et al., 2009; Nakamura et al., 2009; Koretsune and Hotta, 2014). In the quantum spin liquid phase, several exotic properties are observed, such as the (marginally) gapless nature of low-temperature excitations (Shimizu et al., 2003; Yamashita, S., et al., 2008; Yamashita, M., et al., 2009) and emergent inhomogeneity under a magnetic field (Shimizu et al., 2006). It is noted that the slope of the first-order transition line is determined by the difference of entropy between the insulating and metallic phases, obeying the Clausius–Clapeyron relation. The bend of the transition line above the Néel temperatures for κ-Cl indicates a rapid decrease in spin entropy of the insulating phase due to antiferromagnetic short-range order before the long-range order.

Such a re-entrant first-order transition line is reproduced by the CDMFT studies (Park et al., 2008; Liebsch et al., 2009). In contrast, almost negligible bending in the transition line for κ-$Cu_2(CN)_3$ is due to the absence of magnetic order. In the metallic phase, superconductivity emerges at approximately 13 K for κ-Cl (Lefebvre et al., 2000; Limelette et al., 2003a; Kagawa et al., 2004) and 4 K for κ-$Cu_2(CN)_3$ (Kurosaki et al., 2005; Kobashi, 2007; Shimizu et al., 2010). As for the family of $Y[Pd(dmit)_2]_2$, this chapter deals with the compound with $Y = EtMe_3Sb$ (EtMe$_3$Sb-dmit, hereafter), which exhibits a Mott transition from a quantum spin liquid to a metal, as in κ-$Cu_2(CN)_3$; however, a clear first-order transition and a superconducting transition appear to be absent (Kato et al., 2007). The critical temperature of EtMe$_3$Sb-dmit, if it is finite, should be much lower than 30 K (Furukawa et al., 2015; Kato et al., 2007).

4.5 Critical Phenomena around the Critical End Point

According to the theory of critical phenomena, in the vicinity of the critical point, the order parameter ϕ obeys the characteristic scaling behavior, such as

$$\phi(T, h) \propto |T - T_c|^\beta \, (T < T_c, h = 0),$$

$$\phi(T, h) \propto |h|^{1/\delta} (T = T_c),$$

$$\partial\phi(T, h)/\partial h \propto |T - T_c|^{-\gamma} \, (h = 0),$$

where h is the conjugate field of the order parameter, and β, δ, and γ are the critical exponents. In the case of the Mott transition, the density of "doublon," which is double occupancy of electrons at each lattice site, is the order parameter, ϕ, according to the early suggestion by Castellani et al. (1979). The deviation of a pressure from the metal–insulator boundary, $P - P_0(T)$, plays a role of conjugate field h. As shown in Fig. 4.7, Mott insulators and metals can be regarded as doublon-poor and doublon-rich phases, respectively. The choice of the order parameter is supported by DMFT, which shows that the doublon density varies singularly at the

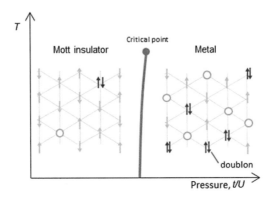

Figure 4.7 Doublon density in the Mott insulating and metallic phases. The doublon density is high in the metallic phase and low in the Mott insulating phase.

critical point (Kotliar et al., 2000). Although doublon density itself is difficult to measure in real experiments, one can exploit physical quantities that are reasonably assumed to be proportional to the doublon density, such as electric conductivity.

In line with the above idea, conductivity measurements around the critical end point have been performed under continuously controlled pressure for Cr-doped V_2O_3 (Limelette et al., 2003b) and for κ-Cl (Kagawa et al., 2005). Both materials showed the scaling of conductivity explained; the case of κ-Cl is shown in Fig. 4.8. The sets of critical exponents deduced are ($\delta \approx 3$, $\beta \approx 0.5$, $\gamma \approx 1$) for Cr-doped V_2O_3 and ($\delta \approx 2$, $\beta \approx 1$, $\gamma \approx 1$) for κ-Cl. As to the Cr-doped V_2O_3, the exponents ($\delta \approx 3$, $\beta \approx 0.5$, $\gamma \approx 1$) are identical to those derived from a mean field theory; furthermore, very near the critical point, it appears that ($\delta \approx 5$, $\beta \approx 0.34$), which are expected for the 3D Ising universality class, better fit the data, being consistent with the theoretical suggestion based on DMFT (Kotliar et al., 2000). In contrast, the exponents for κ-Cl are unexpected; the values of ($\delta \approx 2$, $\beta \approx 1$, $\gamma \approx 1$) do not find experimentally known universality classes.

There are two points of view with respect to the unconventional exponents observed in κ-Cl. One is that the unconventional exponents simply mean that the Mott transition in a quasi-2D system belongs to an unconventional universality class. The other is that the critical exponents are not appropriately extracted because

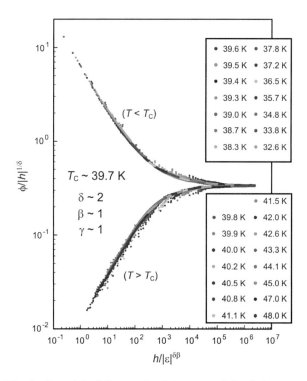

Figure 4.8 Scaling plot of the conductivity around the finite-temperature critical point in κ-Cl (Kagawa et al., 2005).

conductivity is not proportional to the order parameter of the Mott transition (Papanikolaou et al., 2008) or the data points used in the analysis are not sufficiently close to the critical point (Sémon and Tremblay, 2012). Although this issue is still under debate, it is noteworthy that the unconventional exponent $\delta \approx 2$ has also been observed in NMR spin lattice relaxation rate divided by temperature, $(T_1 T)^{-1}$ (Kagawa et al., 2009), as shown in Fig. 4.9, where $(T_1 T)^{-1}$ measures the imaginary part of dynamical spin susceptibility. Assuming that the $(T_1 T)^{-1}$ is proportional to the density of singly occupied sites, whose decrease corresponds to the increase in the doublon density, $(T_1 T)^{-1}$ should indirectly measure the critical variation of the doublon density around the critical end point. The numerical study based on CDMFT also suggests that the exponent δ is much smaller than the expected values for the Ising

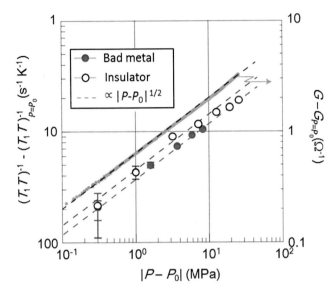

Figure 4.9 Logarithmic plot of $(T_1 T)^{-1}$ and conductance, G, of κ-Cl measured from the critical end point versus the distance from the critical pressure (Kagawa et al., 2009).

universality class in two dimensions ($\delta = 15$) and for the single-site DMFT ($\delta = 3$) (Sentef et al., 2011). A different theoretical argument on the unconventional criticality was made by Imada et al., who focused on the topological nature of the Mott transition, namely the existence or absence of Fermi surfaces, and developed the notion of a marginal quantum criticality for the Mott transition (Misawa and Imada, 2007). The deduced exponents coincide with the observed unconventional ones. On the other hand, there are experimental and theoretical suggestions supporting the second scenario. The scaling analysis of the expansivity of κ-(ET)$_2 X$ pointed to the 2D Ising universality class (Bartosch et al., 2010). Concerning the validity of the conductivity as the order parameter of the Mott transition, Papanicolaou et al. argued that conductivity is not solely determined by the order parameter of the transition but also depends on the singular energy density (Papanikolaou et al., 2008). In addition, a DMFT study suggested that the experimental value of $\delta \approx 2$ is an ostensible value arising from the subleading corrections in the mean-field regime when a system is not so close to the critical point

(Sémon and Tremblay, 2012). Very recently, a report on the critical phenomena in $EtMe_3P[Pd(dmit)_2]_2$ has come out (Abdel-Jawad et al., 2015); the behavior is similar to that of $\kappa\text{-}(ET)_2X$, suggesting the universal nature of the critical phenomena in quasi-2D organic Mott systems.

4.6 Quantum Criticality at Intermediate Temperatures

In this section, we see an experimental observation of the quantum criticality of the Mott transition in organic materials $\kappa\text{-}Cu_2(CN)_3$, κ-Cl, and $EtMe_3Sb$-dmit (Furukawa et al., 2015). As argued in Section 4.2.2, a theoretical study based on DMFT (Terletska et al., 2011; Vučičević et al., 2013) suggests that, even if the critical temperature, T_c, is finite, the quantum criticality can be observed in the intermediate-temperature range, $T_c < T \ll U$, W (Fig. 4.3). Indeed, the critical temperatures of the three compounds are 2 or 3 orders of magnitude lower than the values of W and U, which are several thousand Kelvin or more (Kato and Hengbo, 2012; Komatsu et al., 1996). At low temperatures, the three compounds exhibit different types of ground states across the Mott transitions such as quantum spin liquid (QSL)–superconductor (SC) for κ-$Cu_2(CN)_3$ (Kurosaki et al., 2005; Kobashi, 2007), antiferromagnetic insulator (AFM)–SC for κ-Cl (Lefebvre et al., 2000; Limelette et al., 2003a; Kagawa et al., 2004), and QSL–paramagnetic metal (PM) for $EtMe_3Sb$-dmit (Kato et al., 2007).

4.6.1 Resistivity in the Crossover Region

The quantum criticality of the Mott transition is examined by resistivity measurements under continuously controlled He gas pressure P at various fixed temperatures T above T_c and the scaling analysis following the theoretical suggestion (Terletska et al., 2011; Vučičević et al., 2013) described in Section 4.2.2. Figure 4.10a–c shows the pressure dependences of resistivity $\rho(P, T)$ of κ-$Cu_2(CN)_3$, κ-Cl, and $EtMe_3Sb$-dmit at various temperatures. For all the three materials, the resistivity at low fixed temperatures

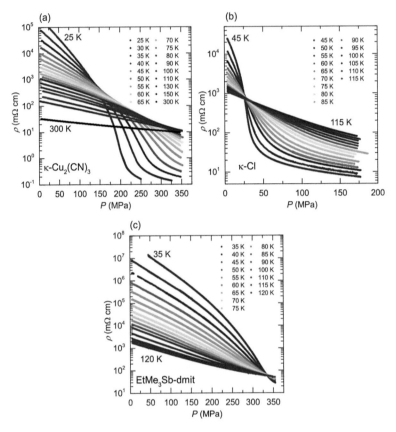

Figure 4.10 Pressure dependence of resistivity for κ-Cu$_2$(CN)$_3$ (a), for κ-Cl (b), and for EtMe$_3$Sb-dmit (c).

steeply decreases by several orders of magnitude as pressure increases, indicating Mott insulator-to-metal crossovers. At higher temperatures, the slopes of the curves become moderate.

To conduct the characteristic scaling analysis, it is prerequisite to determine a metal–insulator crossover line, $P_c(T)$, as the quantum Widom line of the Mott transition in the temperature–pressure phase diagram. The DMFT study (Vučičević et al., 2013) suggests that the quantum Widom line is approximately a locus of the inflection point $P_c(T)$ of the $\log[\rho(P, T)]$ versus P curve at fixed T. This definition makes sense because the obtained $P_c(T)$'s for

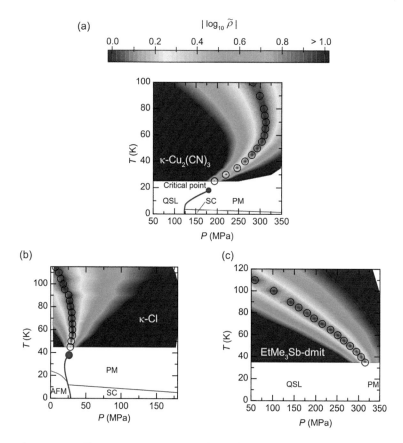

Figure 4.11 Temperature–pressure phase diagrams of κ-Cu$_2$(CN)$_3$ (a), κ-Cl (b), and EtMe$_3$Sb-dmit (c). The open circles indicate the pressure on the quantum Widom line $P_c(T)$. The contour represents the magnitude of $|\log_{10}\tilde{\rho}|$. The lowest temperature experimentally accessible for EtMe$_3$Sb-dmit is limited to 35 K, below which the He medium is solidified at pressures of interest for EtMe$_3$Sb-dmit.

κ-Cu$_2$(CN)$_3$ and κ-Cl appear to be smoothly connected to their low-temperature first-order Mott transition line as shown in Figs. 4.11a and 4.11b. Now that $P_c(T)$ and $\rho_c(T)$ are known, $\rho(P, T)$ in the vertical axis and P in the horizontal axis in Fig. 4.10a–c are converted to the normalized resistivity, $\tilde{\rho} \equiv \rho(P, T)/\rho_c(T)$, and the pressure deviation from the Widom line, $\delta P \equiv P - P_c(T)$, respectively. Figure 4.12 shows $\tilde{\rho}(\delta P, T)$ versus δP for κ-Cu$_2$(CN)$_3$.

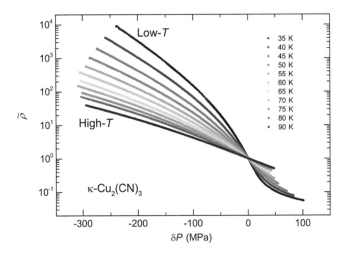

Figure 4.12 Normalized resistivity $\tilde{\rho}(\delta P, T)$ of κ-Cu$_2$(CN)$_3$ versus $\delta P = P - P_c(T)$.

The $\tilde{\rho}(\delta P, T)$ curves construct a data set for testing the quantum-critical scaling.

4.6.2 Scaling Analysis

Before going to the details, the fundamentals of quantum-critical scaling are described. In a quantum-critical regime, both the spatial correlation length ξ and correlation time τ diverge as $\xi \propto |g - g_c|^{-\nu}$ and $\tau \propto \xi^z \propto |g - g_c|^{-z\nu}$ (Sachdev, 2011; Sondhi et al., 1997), where g is the parameter controlling quantum fluctuations, g_c is the critical value of g, ν is the critical exponent of the correlation length, and z is the dynamical exponent. Even in thermal equilibrium, the criticality in dynamics appears inevitably as a characteristic of quantum criticality. The temperature T gives another length scale $L_T \propto T^{-1}$. The L_T is the system size in the temporal direction, and $L_T \to \infty$ when $T \to 0$. The appearance of the L_T indicates that thermal fluctuations disturb the quantum-critical fluctuations slower than the cutoff timescale of L_T. Therefore, the finite-temperature effect can be treated as a kind of finite-size scaling, and physical quantities in the quantum-critical regime at finite temperatures are determined only by the ratio $\tau/L_T \propto T/|g-$

$g_c|^{z\nu}$. In the case of the quantum criticality of the Mott transition, $\delta P = P - P_c(T)$ instead of $P - P_c(T = 0)$ should be used for the scaling analysis according to the bow-shaped crossover line, as shown in Fig. 4.11. If the normalized resistivity follows the scaling, $\tilde{\rho}(\delta P, T) = f(T/|c\delta P|^{z\nu})$, it means that the system is in the quantum-critical regime of the Mott transition, where $f(x)$ is a scaling function and c is an arbitrary constant.

Now, the scaling analysis is performed using the $\tilde{\rho}(\delta P, T)$ data for the three organic compounds of κ-Cu$_2$(CN)$_3$, κ-Cl, and EtMe$_3$Sb-dmit. Figure 4.13a shows the scaling for κ-Cu$_2$(CN)$_3$; $\tilde{\rho}$ is plotted against $T/T_0 = T/|c\delta P|^{z\nu}$ with a critical exponent of $z\nu = 0.62 \pm 0.02$ for $T = 35$–90 K. It is evident that all the data collapse onto two bifurcating scaling curves. It is remarkable that the scaling is satisfied over several orders of magnitude. In a similar fashion, $\tilde{\rho}(\delta P, T)$ of κ-Cl (for 75 K $\leq T \leq 115$ K) and that of EtMe$_3$Sb-dmit (for 35 K $\leq T \leq 90$ K) satisfy the scaling with $z\nu = 0.49 \pm 0.01$ (Fig. 4.13b) and $z\nu = 0.68 \pm 0.04$ (Fig. 4.13c), respectively. Although the three systems have different types of ground states separated by the Mott transitions, namely QSL-SC, AF-SC, and QSL-PM, they exhibit similar scaling features in the intermediate-temperature region, where the present quantum criticality of the Mott transition arises from the high-energy itinerant-localized competition and is reasonably insensitive to the ground states.

Noticeably, the scaling curves of the insulating ($\tilde{\rho} > 1$) and metallic ($\tilde{\rho} < 1$) branches at $T/T_0 > 1$ are highly symmetric, which means that the scaling function $f(x)$ has duality $f_{\text{metal}}(x) = 1/f_{\text{insulator}}(x)$ between the metallic ($\delta P > 0$) and insulating ($\delta P < 0$) sides at $x > 1$. Interestingly, such a mirror symmetry in the scaling function is also found in a metal–insulator transition in the 2D electron gas in a metal-oxide-semiconductor field-effect transistor (MOSFET) (Kravchenko et al., 1995; Simonian et al., 1997), in which a metal–insulator transition is discussed in terms of Wigner crystallization (Amaricci et al., 2010).

The form of the scaling curves indicates the crossover behavior from a high-T/T_0 quantum-critical regime to a low-T/T_0 regime. At $T/T_0 \gg 1$, namely at $|\delta P| \approx 0$, both the metallic and insulating branches are close to each other (Fig. 4.13a–c), meaning large quantum-critical fluctuations between the metallic and insulating

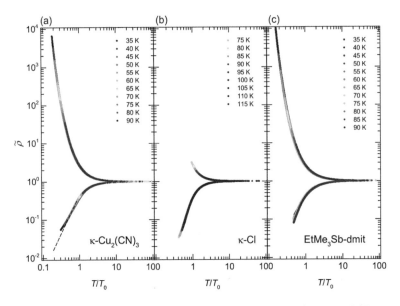

Figure 4.13 Scaling plot of normalized resistivity $\tilde{\rho}(\delta P, T)$ versus $T/T_0 = T/|c\delta P|^{zv}$ with the present values, $zv = 0.62$ and $c = 20.9$ for κ-Cu$_2$(CN)$_3$ (a), $zv = 0.49$ and $c = 289$ for κ-Cl (b), and $zv = 0.68$ and $c = 13.1$ for EtMe$_3$Sb-dmit (c). The upper and lower curves correspond to the insulating and metallic branches, respectively. The dashed line in (a) indicates that $\tilde{\rho} \propto (T/T_0)^2$.

phases. As T/T_0 decreases, (as $|\delta P|$ is increased at fixed T), the bifurcation becomes more obvious, indicating that the system gradually notices the side which it belongs to. To clarify the functional form of the scaling curves, we display the $\tilde{\rho}(\delta P, T)$ versus T/T_0 plot in logarithmic scales for κ-Cu$_2$(CN)$_3$, κ-Cl, and EtMe$_3$Sb-dmit in Fig. 4.14a–c. Further decreases in T/T_0 makes both branches to deviate from the quantum-critical behavior toward the low-temperature behaviors as shown in Fig. 4.14: In the metallic branch, the deviation is from the quantum-critical form $\tilde{\rho} = \exp[-(T/T_0)^{-1/zv}]$ to the Fermi liquid T^2 behavior followed by saturation to the residual resistivity, which is observed especially in κ-Cu$_2$(CN)$_3$ (see also Fig. 4.13a). In the insulating branch, $\tilde{\rho}(\delta P, T)$ departs from the form $\tilde{\rho}(\delta P, T) = \exp[(T/T_0)^{-1/zv}]$ toward near-activation behavior fitted by the form $\tilde{\rho}(\delta P, T) = \exp[\{T/(1.4T_0)\}^{-1.1}]$ for κ-Cu$_2$(CN)$_3$ and

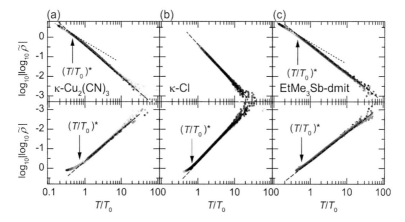

Figure 4.14 Plot of $\log_{10}|\log_{10}(\tilde{\rho})|$ versus T/T_0 for κ-Cu$_2$(CN)$_3$(a), for κ-Cl(b), and for EtMe$_3$Sb-dmit(c). The upper and lower panels are the insulating and metallic branches, respectively. The dashed lines correspond to quantum-critical behavior $\tilde{\rho} = \exp\left[\pm (T/T_0)^{-1/z\nu}\right]$ (+, insulating branch; $-$, metallic branch; $z\nu = 0.62$ for κ-Cu$_2$(CN)$_3$;$z\nu = 0.49$ for κ-Cl;$z\nu = 0.68$ for EtMe$_3$Sb-dmit). The dotted lines correspond to nearly activation-type behavior $\tilde{\rho} = \exp\left[\{T/(1.4T_0)\}^{-1.1}\right]$ for κ-Cu$_2$(CN)$_3$ and $\tilde{\rho} = \exp\left[\{T/(1.2T_0)\}^{-0.9}\right]$ for EtMe$_3$Sb-dmit. The arrows indicate the characteristic value of T/T_0, represented as $(T/T_0)^*$, for the entrance to the low-temperature regimes of the gapped Mott insulator or the Fermi liquid, determined by the eye.

$\tilde{\rho}(\delta P, T) = \exp[\{T/(1.2T_0)\}^{-0.9}]$ for EtMe$_3$Sb-dmit with gaps of the order of T_0's, showing crossovers to gapped Mott insulators.

In order to visualize the quantum-critical region in which the mirror symmetry of the scaling curves is almost perfect, the values of the normalized resistivity $\tilde{\rho}(\delta P, T)$ are represented by contour on the temperature–pressure phase diagram as shown in Fig. 4.11a–c. The fan-shaped region is the quantum-critical region. When the horizontal axis P in the phase diagram is changed to δP, Fig. 4.11a, for example, is reduced to Fig. 4.15, where the crossover temperatures T^* from the quantum-critical state to ground states are also shown. T^* was defined as a temperature at which the T/T_0 dependence of the normalized resistivity changes as shown in Fig. 4.14. The symmetric fan-shaped form of the quantum-critical region against the $\delta P = 0$ is also highlighted in Fig. 4.15.

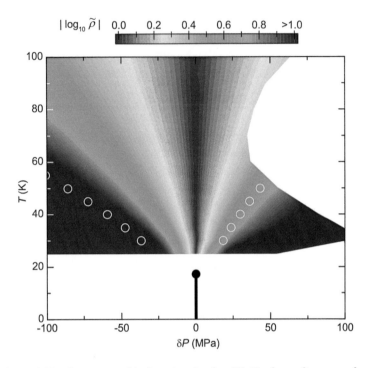

Figure 4.15 Quantum-critical region in the δP–T phase diagram of κ-$Cu_2(CN)_3$. A range of gradation represents the magnitude of $|\log_{10}\tilde{\rho}|$. The insulating and metallic states form at $\delta P < 0$ and $\delta P > 0$, respectively. The open circles indicate the characteristic temperatures, T^*, for the entrance to the low-temperature regimes of the gapped Mott insulator ($\delta P < 0$) or the Fermi liquid ($\delta P > 0$).

4.6.3 Comparison of the Experimental Results with the DMFT Predictions

The experimental results shown above are consistent with the theoretical predictions of DMFT of the Hubbard model (Terletska et al., 2011; Vučičević et al., 2013) in many respects: (i) the scaling curves, (ii) the phase diagrams, and (iii) the critical exponent $z\nu$. First, both experimental and theoretical scaling functions have approximately the same form with the mirror symmetry between metallic and insulating branches, suggesting that the local correlation considered in DMFT plays a primary role in the quantum-critical transport. Besides the form of the scaling curves, the experimental and theoretical phase diagrams resemble each other, in particular for κ-$Cu_2(CN)_3$. Both experimental and theoretical phase diagrams have a fan-shaped quantum-critical region and a bow-shaped quantum Widom line smoothly connected with the low-temperature first-order transition line. Furthermore, the critical exponent $z\nu = 0.56$–0.57 derived from DMFT (Terletska et al., 2011) is close to the experimental values $z\nu = 0.62$ for κ-$Cu_2(CN)_3$, 0.49 for κ-Cl, and 0.68 for $EtMe_3Sb$-dmit. Interestingly, the quantum field theory that argues the genuine quantum Mott transition in a 2D system also yields the critical exponent $z\nu = 0.67$ (Senthil, 2008; Witczak-Krempa et al., 2012), which is comparable to the experimental values. The temperatures at which the normalized resistivity satisfies the quantum-critical scaling approximately range from $2T_c$ to 0.1–$0.2t$, where t is the transfer integral between nearest-neighbor sites. These lower and upper limits for energies pertinent to the quantum criticality are nearly identical to those predicted by DMFT (Terletska et al., 2011). In the (single-site) DMFT framework, k-dependent fluctuations are neglected; thus, the above agreements between the experimental and DMFT results suggest that the present quantum-critical fluctuations may be local in nature.

The values of the critical exponent $z\nu$ for the three organic materials ($z\nu = 0.62 \pm 0.02$, 0.49 ± 0.02 and 0.68 ± 0.04) are appreciably different but, on the other hand, reside in a narrow window of 0.6 ± 0.1. It is conceivable that the present systems belong to the same universality class with $z\nu \approx 0.6$ because the scattering in the obtained $z\nu$ values is much smaller than the range of the known $z\nu$ values for quantum metal–insulator transitions; for instance, $z\nu = 2$ ($z = 4$, $\nu = 1/2$) is proposed theoretically for the marginal quantum Mott transition in (Misawa and Imada, 2007) and $z\nu = 1.6$ ($z = 1$, $\nu = 1.6$) is found experimentally in a Si-MOSFET by Kravchenko et al. (1995). Another possible viewpoint is that the small but appreciable variation in the $z\nu$ values is meaningful and signifies nonuniversal critical exponents like those of the Berezinskii–Kosterlitz–Thouless transition (Kosterlitz, 1974). In this case, $z\nu$ depends on the microscopic parameters of a system such as lattice geometry.

4.7 Summary

In this chapter, the recent experimental studies on the Mott transitions of layered organic materials κ-$(ET)_2 X$ and $Y[Pd(dmit)_2]_2$ are reviewed with an emphasis on the quantum-critical transport. In the vicinity of the Mott transition, at temperatures sufficiently lower than the critical point, different kinds of phases emerge, such as antiferromagnetic Mott insulators, quantum spin liquids, Fermi liquids, and unconventional superconductors. Around the finite-temperature critical end point, the critical phenomena associated with the Mott transition can be observed. On top of that, experimental studies in the crossover region well above the critical point reveal the quantum criticality of the Mott transition. The observed quantum criticality shows material-independent nature and is consistent with the DMFT predictions. Interactions among electrons provide diverse electronic phases at low energies; however, the interactions drive the systems near the metal–insulator transition into a material-independent quantum-critical state. The electronic quantum-critical fluid can be novel electronic states, clarification of which is a forthcoming issue.

References

Abdel-Jawad, M., Kato, R., Watanabe, I., Tajima, N., and Ishii, Y. (2015). *Phys. Rev. Lett.* **114**, p. 106401.

Amaricci, A., et al. (2010). *Phys. Rev. B* **82**, p. 155102.

Bartosch, L., de Souza, M., and Lang, M. (2010). *Phys. Rev. Lett.* **104**, p. 245701.

Brinkman, W. F., and Rice, T. M. (1970). *Phys. Rev. B* **2**, pp. 4302–4304.

Castellani, C., Castro, C. D., Feinberg, D., and Ranninger, J. (1979). *Phys. Rev. Lett.* **43**, pp. 1957–1960.

De Souza, M., et al. (2007). *Phys. Rev. Lett.* **99**, p. 037003.

Fournier, D., Poirier, M., Castonguay, M., and Truong, K. D. (2003). *Phys. Rev. Lett.* **90**, p. 127002.

Furukawa, T., Miyagawa, K., Taniguchi, H., Kato, R., and Kanoda, K. (2015). *Nat. Phys.* **11**, pp. 221–224.

Georges, A., Kotliar, G., Krauth, W., and Rozenberg, M. J. (1996). *Rev. Mod. Phys.* **68**, pp. 13–125.

Hubbard, J. (1964). *Proc. R. Soc. A Math. Phys. Eng. Sci.* **281**, pp. 401–419.

Imada, M., Fujimori, A., and Tokura, Y. (1998). *Rev. Mod. Phys.* **70**, pp. 1039–1263.

Kagawa, F., Itou, T., Miyagawa, K., and Kanoda, K. (2004). *Phys. Rev. B* **69**, p. 064511.

Kagawa, F., Miyagawa, K., and Kanoda, K. (2005). *Nature* **436**, pp. 534–537.

Kagawa, F., Miyagawa, K., and Kanoda, K. (2009). *Nat. Phys.* **5**, pp. 880–884.

Kandpal, H., Opahle, I., Zhang, Y.-Z., Jeschke, H. O., and Valentí, R. (2009). *Phys. Rev. Lett.* **103**, p. 067004.

Kanoda, K. (1997a). *Physica C* **282–287**, pp. 299–302.

Kanoda, K. (1997b). *Hyperfine Interact.* **104**, pp. 235–249.

Kanoda, K. (2006). *J. Phys. Soc. Jpn.* **75**, p. 051007.

Kanoda, K., and Kato, R. (2011). *Annu. Rev. Condens. Matter Phys.* **2**, pp. 167–188.

Kato, R. (2014). *Bull. Chem. Soc. Jpn.* **87**, pp. 355–374.

Kato, R., and Hengbo, C. (2012). *Crystals* **2**, pp. 861–874.

Kato, R., Tajima, A., Nakao, A., Tajima, N., and Tamura, M. (2007). *Multifunctional Conducting Molecular Materials*, pp. 32–38, eds. G. Saito et al. (Royal Society of Chemistry).

Kino, H., and Fukuyama, H. (1995). *J. Phys. Soc. Jpn.* **64**, pp. 2726–2729.

Kino, H., and Fukuyama, H. (1996). *J. Phys. Soc. Jpn.* **65**, pp. 2158–2169.

Kobashi, K. (2007). *Transport Properties Near the Mott Transition in the Quasi-Two-Dimensional Organic Conductor κ-(ET)2X*. Thesis, University of Tokyo.

Komatsu, T., Matsukawa, N., Inoue, T., and Saito, G. (1996). *J. Phys. Soc. Jpn.* **65**, pp. 1340–1354.

Koretsune, T., and Hotta, C. (2014). *Phys. Rev. B* **89**, p. 045102.

Kosterlitz, J. M. (1974). *J. Phys. C Solid State Phys.* **7**, pp. 1046–1060.

Kotliar, G., Lange, E., and Rozenberg, M. J. (2000). *Phys. Rev. Lett.* **84**, pp. 5180–5183.

Kravchenko, S. V., et al. (1995). *Phys. Rev. B* **51**, pp. 7038–7045.

Kurosaki, Y., Shimizu, Y., Miyagawa, K., Kanoda, K., and Saito, G. (2005). *Phys. Rev. Lett.* **95**, p. 177001.

Lefebvre, S., et al. (2000). *Phys. Rev. Lett.* **85**, pp. 5420–5423.

Liebsch, A., Ishida, H., andMerino, J. (2009). *Phys. Rev. B* **79**, p. 195108-1.

Limelette, P., et al. (2003a). *Phys. Rev. Lett.* **91**, p. 016401.

Limelette, P., et al. (2003b). *Science* **302**, pp. 89–92.

McWhan, D. B., Menth, A., Remeika, J. P., Brinkman, W. F., and Rice, T. M. (1973). *Phys. Rev. B* **7**, pp. 1920–1931.

Misawa, T., and Imada, M. (2007). *Phys. Rev. B* **75**, p. 115121.

Miyagawa, K., Kawamoto, A., Nakazawa, Y., and Kanoda, K. (1995). *Phys. Rev. Lett.* **75**, pp. 1174–1177.

Mori, T., et al. (1984). *Bull. Chem. Soc. Jpn.* **57**, pp. 627–633.

Mott, N. F. (1990). *Metal-Insulator Transitions* (Taylor & Francis).

Nakamura, K., Yoshimoto, Y., Kosugi, T., Arita, R., and Imada, M. (2009). *J. Phys. Soc. Jpn.* **78**, p. 083710.

Papanikolaou, S., et al. (2008). *Phys. Rev. Lett.* **100**, p. 026408.

Parcollet, O., Biroli, G., and Kotliar, G. (2004). *Phys. Rev. Lett.* **92**, pp. 226402-1.

Park, H., Haule, K., and Kotliar, G. (2008). *Phys. Rev. Lett.* **101**, p. 186403-1.

Sémon, P., and Tremblay, A.-M. S. (2012). *Phys. Rev. B* **85**, p. 201101.

Sachdev, S. (2011). *Quantum Phase Transitions* (Cambridge University Press).

Sentef, M.,Werner, P., Gull, E., and Kampf, A. P. (2011). *Phys. Rev. B* **84**, p. 165133.

Senthil, T. (2008). *Phys. Rev. B* **78**, p. 045109.

Shimizu, Y., et al. (2010). *Phys. Rev. B* **81**, p. 224508.

Shimizu, Y., Miyagawa, K., Kanoda, K., Maesato, M., and Saito, G. (2003). *Rev. Lett.* **91**, p. 107001.

Shimizu, Y., Miyagawa, K., Kanoda, K., Maesato, M., and Saito, G. (2006). *Phys. Rev. B* **73**, p. 140407.

Simonian, D., Kravchenko, S. V., and Sarachik, M. P. (1997). *Phys. Rev. B* **55**, pp. R13421–R13423.

Sondhi, S. L., Girvin, S. M., Carini, J. P., and Shahar, D. (1997). *Rev. Mod. Phys.* **69**, pp. 315–333.

Terletska, H., Vučičević, J., Tanasković, D., and Dobrosavljević, V. (2011). *Phys. Rev. Lett.* **107**, p. 026401.

Vučičević, J., Terletska, H., Tanasković, D., and Dobrosavljević, V. (2013). *Phys. Rev. B* **88**, p. 075143.

Witczak-Krempa, W., Ghaemi, P., Senthil, T., and Kim, Y. B. (2012). *Phys. Rev. B* **86**, p. 245102.

Yamashita, M., et al. (2009). *Nat. Phys.* **5**, pp. 44–47.

Yamashita, S., et al. (2008). *Nat. Phys.* **4**, pp. 459–462.

Yao, X., Honig, J. M., Hogan, T., Kannewurf, C., and Spałek, J. (1996). *Phys. Rev. B* **54**, pp. 17469–17475.

Zhang, X. Y., Rozenberg, M. J., and Kotliar, G. (1993). *Phys. Rev. Lett.* **70**, pp. 1666–1669.

Chapter 5

Metal–Insulator Transition in Correlated Two-Dimensional Systems with Disorder

Dragana Popović

National High Magnetic Field Laboratory, Florida State University, Tallahassee, Florida 32310, USA
dragana@magnet.fsu.edu

Experimental evidence for the possible universality classes of the metal–insulator transition (MIT) in two dimensions (2D) is discussed. Sufficiently strong disorder, in particular, changes the nature of the transition. Comprehensive studies of the charge dynamics are also reviewed, describing evidence that the MIT in a 2D electron system in silicon should be viewed as the melting of the Coulomb glass. Comparisons are made to recent results on novel 2D materials and quasi-2D strongly correlated systems, such as cuprates.

5.1 2D Metal–Insulator Transition as a Quantum Phase Transition

The metal–insulator transition (MIT) in 2D systems remains one of the most fundamental open problems in condensed matter physics

Strongly Correlated Electrons in Two Dimensions
Edited by Sergey Kravchenko
Copyright © 2017 Pan Stanford Publishing Pte. Ltd.
ISBN 978-981-4745-37-6 (Hardcover), 978-981-4745-38-3 (eBook)
www.panstanford.com

(Abrahams et al., 2001; Dobrosavljević, 2012; Kravchenko and Sarachik, 2004; Spivak et al., 2010). The very existence of the metal and the MIT in 2D had been questioned for many years, but recently, considerable experimental evidence has become available in favor of such a transition. Indeed, in the presence of electron–electron interactions, the existence of the 2D MIT does not contradict any general idea or principle (see also chapters by V. Dobrosavljević, and A. A. Shashkin and S. V. Kravchenko). It is important to recall that a qualitative distinction between a metal and an insulator exists only at temperature $T = 0$: the conductivity $\sigma(T = 0) \neq 0$ in the metal and $\sigma(T = 0) = 0$ in the insulator. Therefore, the MIT is an example of a quantum phase transition (QPT) (Sachdev, 2011): it is a continuous phase transition that occurs at $T = 0$, that is, between two ground states. It is controlled by some parameter of the Hamiltonian of the system, such as carrier density, the external magnetic field, or pressure, and quantum fluctuations dominate the critical behavior. In analogy to thermal phase transitions, a QPT is characterized by a correlation length $\xi \propto |\delta|^{-\nu}$ and the corresponding timescale $\tau_\xi \approx \xi^z$, both of which diverge in a power-law fashion at the critical point. Here δ is a dimensionless (reduced) distance of a control parameter from its critical value, ν is the correlation length exponent, and z is the dynamical exponent. Because of the Heisenberg uncertainty principle, in a QPT there is a characteristic energy (temperature) scale

$$T_0 \approx \frac{\hbar}{\tau_\xi} \approx |\delta|^{\nu z}, \tag{5.1}$$

which decreases to zero in a power-law fashion as the critical point is approached (Fig. 5.1).

The existence of a diverging correlation length leads to the expectation that in the vicinity of a critical point, the behavior of the system will exhibit a certain degree of universality, independent of microscopic details. Indeed, while microscopic theories describing the MIT remain controversial, scaling behavior is a much more robust and general property of second-order phase transitions (e.g., Dobrosavljević, 2012). In the case of an MIT, which is controlled by changing the carrier density n, it is the conductivity that can be described in the critical region (Fig. 5.1) by a scale-invariant form as

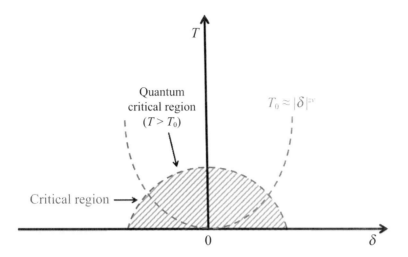

Figure 5.1 The phase diagram of a system near a QPT. A quantum-critical point at $T = 0$ and $\delta = 0$ separates two ground states, for example, an insulator and a metal. The red dashed curve denotes a smooth crossover at $T_0 \approx |\delta|^{\nu z}$. Well-defined differences between the two phases, such as insulating versus metallic behavior, are observed only at $T < T_0$. At $T > T_0$, the system is in the quantum-critical region, in which both quantum and thermal fluctuations are important. Scaling behavior is expected in the entire (blue hatched) critical region.

$$\sigma(n, T) = \sigma_c(T) f(T/T_0) = \sigma_c(T) f(T/|\delta|^{zv}), \qquad (5.2)$$

where $\delta \equiv \delta_n = (n - n_c)/n_c$ (n_c is the critical density) and

$$\sigma_c(T) \equiv \sigma(n = n_c, T) \propto T^x \qquad (5.3)$$

is the critical conductivity. From the scaling description, it also follows that in the metallic phase at $T = 0$,

$$\sigma(n, T = 0) \propto \delta_n^\mu, \qquad (5.4)$$

such that the conductivity exponent $\mu = x(zv)$. Thus the behavior observed in the $T = 0$ limit (Eq. 5.4) also provides a stringent test of the scaling behavior observed at $T \neq 0$ (Eqs. 5.2 and 5.3). One of the goals of the scaling analysis is to reveal critical exponents that characterize the transition and correspond to specific universality classes.

The simplest scaling scenario predicts the validity of Wegner scaling for which the conductivity exponent $\mu = (D - 2)\nu$, where D is the dimensionality of the system. This would imply that at the 2D MIT, $x = (D - 2)/z = 0$, so the critical conductivity should not depend on T. It is for this reason that many early studies of the 2D MIT identified n_c as the carrier density where $d\sigma/dT$ changes sign from insulator-like ($d\sigma/dT > 0$) to metallic ($d\sigma/dT < 0$). However, it should be emphasized that scaling with $x \neq 0$ for 2D systems does not contradict any fundamental principle (Belitz and Kirkpatrick, 1994). Indeed, such violations of Wegner scaling were predicted in the presence of "dangerously irrelevant operators" (Amit and Peliti, 1982), for example, for certain microscopic models with strong spin-dependent components of the Coulomb interactions (Belitz and Kirkpatrick, 1995; Castellani et al., 1987; Kirkpatrick and Belitz, 1994). As discussed below, it is precisely the general scaling form, Eq. 5.2 with $x \neq 0$, that provides a satisfactory and consistent description of all the data near a 2D MIT.

In general, a QPT can affect the behavior of the system up to surprisingly high temperatures. In fact, many unusual properties of various strongly correlated materials have been attributed to the proximity of quantum-critical points. An experimental signature of a QPT at nonzero T is the observation of scaling behavior with relevant parameters in describing the data. At $T \neq 0$, however, the correlation length is finite, so single-parameter scaling in Eq. 5.2 will work only if the sample size $L > \xi$. Otherwise, the scaling function will depend not only on T/T_0, but also on another scaling variable $\sim L/\xi$. Such finite-size effects have not been observed in the experimental studies of the 2D MIT so far: the scaling functions have been found to depend only on T/T_0, indicating that sample sizes have been sufficiently large for the experimental temperature range.

In real systems near an MIT, both disorder and electron–electron interactions may play an important role and lead to out-of-equilibrium or glassy behavior of electrons (e.g., Popović, 2012). The goal of this chapter is thus twofold. First, it will focus on studies of the critical behavior of conductivity in 2D electron systems (2DESs) with different amounts and types of disorder, as well as different ranges of Coulomb interactions, in order to identify possible universality classes of the 2D MIT. Second, experimental results obtained using a variety of protocols to probe charge dynamics will be reviewed, providing important information about the nature of the insulating phase and the 2D MIT. The focus will be on detailed and comprehensive studies that have been carried out so far only on a 2DES in (100)-Si metal-oxide-semiconductor field-effect transistors (MOSFETs). However, recent results on novel 2D systems, in particular, single- or few-layer transition metal dichalcogenides, and on quasi-2D strongly correlated materials, such as cuprates, will be also discussed.

5.2 Critical Behavior of Conductivity

The 2DESs in semiconductor heterostructures (Ando et al., 1982) are relatively simple systems for exploring the interplay of electronic correlations and disorder, because all the relevant parameters—carrier density n_s, disorder, and interactions—can be varied easily. For example, n_s can be tuned over 2 orders of magnitude by applying voltage V_g to the gate electrode. At low n_s, the 2DES is strongly correlated: $r_s \gg 1$, where $r_s = E_C/E_F \propto n_s^{-1/2}$ (E_C is the average Coulomb energy per electron and E_F is the Fermi energy; see also (Kravchenko and Sarachik, 2004) for more details). Therefore, the effects of interactions become increasingly important as n_s is reduced.

In Si-MOSFETs, random potential (disorder) that is "felt" by the 2DES is caused by charged impurities (Na$^+$ ions), which are randomly distributed in the oxide and thus spatially separated from the 2DES. In these devices, the (Drude) mobility $\mu = \sigma/(n_s e)$ of the 2DES peaks as a function of n_s because, at very high n_s that are not of interest here, the scattering due to the roughness of the Si–SiO$_2$ interface becomes dominant (Ando et al., 1982). The peak mobility at 4.2 K, μ_{peak}, is commonly used as a rough measure of the amount of disorder. As described below, detailed studies of $\sigma(n_s, T)$ demonstrate scaling behavior consistent with the existence of a QPT in all 2DESs in Si regardless of the amount of disorder.

5.2.1 Role of Disorder

5.2.1.1 Low-disorder samples

Si-MOSFETs with relatively little disorder ($\mu_{peak}[m^2/Vs] \approx 1 - 3$) make it possible to access lower electron densities, and thus the regime of stronger interactions, before strong localization sets in.[a] The corresponding values of r_s near the critical density have been $r_s > 13$. In such low-disorder samples, the most obvious feature that suggests the existence of a metal is the large, almost an order-of-magnitude increase of conductivity with decreasing temperature ($d\sigma/dT < 0$) observed at[b] $T < T_F$ and low n_s, where $r_s \gg 1$. However, this does *not* necessarily imply $\sigma(T = 0) \neq 0$, that is, a metallic ground state. Therefore, even though such a strong $d\sigma/dT < 0$ had been known for a long time (e.g., Smith and Stiles, 1986), it was not considered as evidence for a metallic state. Indeed, it was only after dynamical scaling (Eq. 5.2) was demonstrated (Kravchenko et al., 1995) and independently confirmed on a different set of devices (Popović et al., 1997) that the problem of the 2D MIT attracted renewed attention.

In those experiments, it was possible to collapse all the $\sigma(n_s, T)$ data near n_c onto the same function $f(T/T_0)$ with two branches, the upper one for the metallic side of the transition and the lower one for the insulating side (see Fig. 5.2), using the same values $zv = 1.6$ and $x = 0$. The temperature ranges for scaling were also comparable: $0.08 \lesssim T/T_F \lesssim 0.4$ in Fig. 5.2 (Popović et al., 1997) and $0.05 \lesssim T/T_F \lesssim 0.3$ in Kravchenko et al. (1995). The critical density n_c was identified as the *separatrix*, that is, the density n_s^* where $d\sigma/dT$ changes sign. Subsequently, other, more appropriate methods were used to identify n_c. In particular, n_c was determined on the basis of both a vanishing activation energy (Jaroszyński et al., 2002, 2004; Shashkin et al., 2001) and a vanishing nonlinearity of current–voltage ($I–V$) characteristics when extrapolated from the insulating phase (Pudalov et al., 1993; Shashkin et al., 2001), and established that indeed $n_c \approx n_s^*$. It is important to stress, however, that this

---

[a]In the strongly localized or insulating regime, the conductivity decreases exponentially with decreasing temperature, leading to $\sigma(T = 0) = 0$.
[b]The Fermi temperature $T_F[K] = 7.31 n_s[10^{11}$ cm$^{-2}]$ for electrons in Si-MOSFETs (Ando et al., 1982).

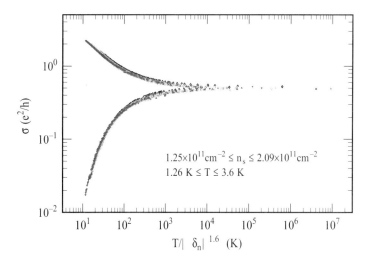

Figure 5.2 Scaling of conductivity in a low-disorder Si-MOSFET with $\mu_{\text{peak}} \approx 1\ \text{m}^2/\text{Vs}$. The scaling ranges of n_s and T are given on the plot; $n_c = (1.67 \pm 0.02) \times 10^{11}\ \text{cm}^{-2}$. Reprinted (figure) with permission from [Popović, D., Fowler, A. B., and Washburn, S. (1997). *Phys. Rev. Lett.* **79**, p. 1543.] Copyright (1997) by the American Physical Society.

is true only in low-disorder samples with nonmagnetic scattering. The situation is very different in the presence of scattering by local magnetic moments and in highly disordered samples. In those cases, the two densities can be vastly different, as shown in Sections 5.2.1.2 and 5.2.1.3.

Therefore, the first studies of scaling near a 2D MIT seemed to be consistent with Wegner scaling, in which the critical conductivity (Eq. 5.3) does not depend on temperature, that is, the exponent $x = 0$. Not long after, it was realized that this simple scaling scenario fails in the metallic regime at the lowest temperatures, where, following its initial large increase, $\sigma(T)$ becomes a weak function as $T \rightarrow 0$ (Altshuler et al., 2001; Pudalov et al., 1998). A careful analysis of the data showed (Washburn et al. 1999a, 1999b) that they could be scaled according to the more general form (Eq. 5.2) with $\sigma_c = \sigma(n_c, T) \propto T^x$, $x \neq 0$ (the scaling range was $0.02 \lesssim T/T_F \lesssim 0.2$). This type of scaling implies, of course, that σ_c vanishes as $T \rightarrow 0$, that is, n_c does not coincide with the separatrix n_s^* but instead $\sigma_c(T)$ belongs to the insulating family of curves. This conclusion is consistent with the results of activation energy and nonlinear $I-V$ studies that reported a small but systematic difference of a few percentage points such that $n_c < n_s^*$ (Altshuler et al., 2001; Pudalov et al., 1998). Nevertheless, it has proven difficult to determine the exponent x reliably, because the weak, low-T behavior of $\sigma(T)$ is observable only over a narrow range of very low temperatures ($T \approx 0.1$ K), when it becomes difficult to cool the electrons in the experiment. In samples that are slightly more disordered ($\mu_{\text{peak}} \lesssim 1$ m^2/Vs), the onset of the weak $\sigma(T)$ (or an apparent saturation on a logarithmic scale) in the metallic regime takes place at higher temperatures, for example, at $T \lesssim 1$ K in Fig. 5.3(top), where it is easy to cool the electrons. This shows that the weak metallic $\sigma(T)$ (saturation) observed at low T is an intrinsic effect and its onset depends on disorder. Here it is still possible to scale the data at $T > 1$ K with $z\nu = 1.6$ and $x = 0$ around the separatrix (open black triangles in Fig. 5.3(top) corresponding to $n_s^* = 1.25 \times 10^{11}$ cm^{-2}), but it is obvious that this curve acquires an insulating T dependence at $T < 1$ K, so Wegner scaling will fail. However, it is still difficult to make reliable fits to $\sigma(T)$ in the low-T regime, so the direct determination of the exponent x in low-disorder samples remains a challenge.

On the other hand, by extrapolating the metallic saturation of σ to $T = 0$, it has been found (Fletcher et al., 2001) that $\sigma(n_s, T = 0) \propto \delta_n^\mu$, with the exponent $\mu = 1$–1.5. The critical density n_c obtained in this way was by about 1%–5% lower than n_s^*, in agreement with the results of the activation energy and nonlinear $I-V$ studies. Some estimates of the exponent x may be made then on the basis of the relation $\mu = x(z\nu)$.

Scaling with $x \neq 0$ is also observed in 3D materials near the MIT (Sarachik, 1995). Other striking similarities near the MIT between transport properties of a 2DES in low-disorder Si-MOSFETs and those observed in Si:B, a 3D system, were pointed out early on (Popović et al., 1997). For example, in Si:B, magnetoconductance, which is negative, depends strongly on the carrier concentration and shows a dramatic decrease at the transition (Bogdanovich et al., 1997). Such anomalous behavior was attributed to electron–electron interactions. In a 2DES, the negative component of the magnetoconductance, which arises from the spin-dependent part of the electron–electron interaction (Lee and Ramakrishnan, 1985),

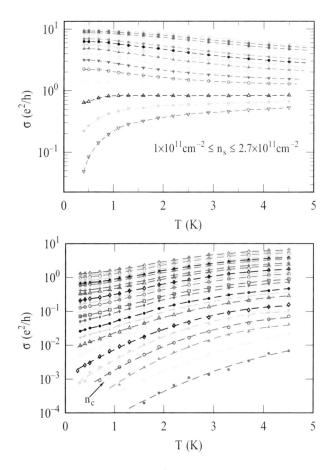

Figure 5.3 Conductivity σ versus T for different n_s in a low-disorder 2DES with $\mu_{\text{peak}} \approx 0.5$–$1$ m^2/Vs. Top: The substrate (back-gate) bias on the sample was $V_{\text{sub}} = -40$ V. The data can be scaled with $x = 0$ (similarly to Fig. 5.2) around the separatrix (open black triangles) $n_s^* = 1.25 \times 10^{11}$ cm^{-2} only at $T > 1$ K. Reprinted (figure) with permission from [Feng, X. G., Popović, D., and Washburn, S. (1999). *Phys. Rev. Lett.* **83**, p. 368.] Copyright (1999) by the American Physical Society. Bottom: $V_{\text{sub}} = +1$ V, which induces scattering by local magnetic moments; $0.3 \leq n_s(10^{11}$ cm$^{-2}) \leq 3.0$. The data can be scaled (Fig. 5.4) around $n_c = 0.55 \times 10^{11}$ cm^{-2} with $x \neq 0$ down to the lowest T. Reprinted (figure) with permission from [Feng, X. G., Popović, D., Washburn, S., and Dobrosavljević, V. (2001). *Phys. Rev. Lett.* **86**, p. 2625.] Copyright (2001) by the American Physical Society.

also decreases rapidly at the transition, suggesting that electron–electron interactions may play a similar role in both 2D and 3D systems near the MIT (Popović et al., 1997).

In recent years, many different types of experiments, including both transport and thermodynamic measurements, have been performed on low-disorder samples. Some of those studies are discussed in detail in the chapter by A. A. Shashkin and S. V. Kravchenko. There is mounting evidence that suggests that electron–electron interactions are responsible for a variety of phenomena observed in the metallic regime of a low-disorder 2DES near the MIT, including a large increase of σ with decreasing temperature ($d\sigma/dT < 0$) (Radonjić et al., 2012). In particular, since the most striking experimental features are not sensitive to weak disorder (see, for example, the thermopower study by Mokashi et al., 2012), they have been interpreted as evidence that the MIT in such low-disorder systems is driven by electron–electron interactions and that disorder has only a minor effect. It is thus important to compare the properties of the MIT, such as critical exponents, in low-disorder samples to those obtained in samples with a different type or amount of disorder.

5.2.1.2 Special disorder: local magnetic moments

It turns out that the metallic behavior with $d\sigma/dT < 0$ is easily suppressed by scattering of the conduction electrons by disorder-induced local magnetic moments. More precisely, even an arbitrarily small amount of such scattering is sufficient to suppress the $d\sigma/dT < 0$ behavior in the $T \to 0$ limit (Feng et al., 1999). It is important to note that this does not necessarily indicate the destruction of the metallic phase. In disordered 3D metals, for example, it is well known that the derivative $d\sigma/dT$ can be either negative or positive near the MIT (Lee and Ramakrishnan, 1985). Indeed, as described below, in the presence of local magnetic moments, the 2DES exhibits in the metallic phase the simplest $\sigma(T)$, observed over two decades of T, allowing for an unambiguous extrapolation to $T = 0$ and an excellent fit to the dynamical scaling described by Eq. 5.2.

In Si-MOSFETs, it is possible to change the disorder, for a fixed n_s, by applying bias V_{sub} to the Si substrate (back gate) (Ando et al., 1982). In particular, the reverse (negative) V_{sub} moves the electrons closer to the interface, which increases the disorder. It also increases the splitting between the subbands since the width of the triangular potential well at the interface in which the electrons are confined is reduced by applying negative V_{sub}. Usually, only the lowest subband is occupied at low T, giving rise to the 2D behavior. However, in sufficiently disordered samples or wide enough potential wells, the band tails associated with the upper subbands can be so long that some of their strongly localized states may be populated even at low n_s, and act as additional scattering centers for 2D electrons. In particular, since at least some of them must be singly populated due to a large on-site Coulomb repulsion (tens of milli–electron volts), they may act as local magnetic moments. Clearly, the negative V_{sub} reduces this type of scattering by depopulating the upper subband. Therefore, the effects of local magnetic moments on the transport properties of the 2DES were studied by varying V_{sub} (Feng et al., 1999, 2001). It is important to note that the presence of local moments did not have a significant effect on the value of μ_{peak}, probably because the upper subband is depopulated at high V_g, where mobility peaks.[c]

Figure 5.3 illustrates the effect of local magnetic moments on $\sigma(T)$ in the same sample: while $d\sigma/dT < 0$ is observed at high n_s in the absence of local moments (top panel), $\sigma(T)$ curves become insulator-like $(d\sigma/dT > 0)$ for all n_s after many local moments are introduced (bottom panel). In the latter case, the critical density can be easily identified by plotting the data on a log-log scale (not shown): at n_c, the conductivity obeys a pure power-law form (Eq. 5.3) with $x \approx 2.6$ (Feng et al., 2001). For $n_s < n_c$, $\sigma(T)$ decreases exponentially, as expected in the insulating phase. For $n_s > n_c$, a simple and precise form $\sigma(n_s, T) = \sigma(n_s, T = 0) + A(n_s)T^2$ is observed from about 2 K down to the lowest accessible $T = 0.020$ K (Eng et al., 2002). Figure 5.4 (top) shows that the extrapolated zero-temperature conductivity is a power-law function of δ_n (Eq. 5.4), that is, $\sigma(n_s, T = 0) \propto \delta_n^\mu$ ($\mu \approx 3$), as expected in the vicinity of

---

[c]For a fixed V_{sub}, the subband splitting increases by increasing V_g (Ando et al., 1982).

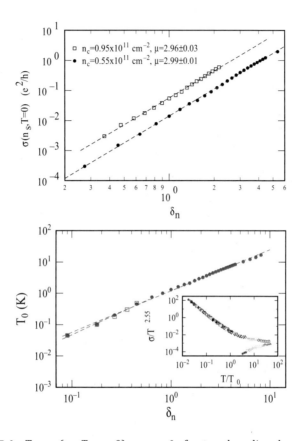

Figure 5.4 Top: $\sigma(n_s, T = 0)$ versus δ_n for two low-disorder samples with different n_c, as shown, in the presence of scattering by local magnetic moments. The dashed lines are fits with the slopes equal to the critical exponent μ. At the MIT, the corresponding $r_s \approx 17$ and 22 for the two samples, respectively. Bottom: Scaling parameter T_0 as a function of $|\delta_n|$ for the sample with $n_c = 0.55 \times 10^{11} \mathrm{cm}^{-2}$ (Fig. 5.3, bottom); open symbols: $n_s < n_c$, closed symbols: $n_s > n_c$. The dashed lines are fits with slopes $z\nu = 1.4 \pm 0.1$ and $z\nu = 1.32 \pm 0.01$, respectively. Inset: Scaling of raw data $\sigma/\sigma_c \approx \sigma/T^x$ in units of $e^2/hK^{2.55}$ for all n_s shown in Fig. 5.3 (bottom) and $T < 2$ K. Reprinted (figure) with permission from [Feng, X. G., Popović, D., Washburn, S., and Dobrosavljević, V. (2001). *Phys. Rev. Lett.* **86**, p. 2625.] Copyright (2001) by the American Physical Society.

a quantum-critical point (Goldenfeld, 1992). Furthermore, the data obey dynamical scaling (Eq. 5.2) with $z\nu \approx 1.3$ (Fig. 5.4, bottom) and a T-dependent critical conductivity (Eq. 5.3) with $x \approx 2.6$, that is, in agreement with theoretical expectations near a QPT. The consistency of the scaling analysis is confirmed by comparing $\mu = x(z\nu) = 3.4 \pm 0.4$ obtained from scaling with $\mu = 3.0 \pm 0.1$ determined from the $T = 0$ extrapolation of $\sigma(n_s, T)$ (Fig. 5.4, top).

The exponent $z\nu$ is thus comparable to that found in low-disorder samples in the absence of scattering by local magnetic moments (Section 5.2.1.1; see also Section 5.2.4). On the other hand, unlike $z\nu$, the exponent x seems to be more sensitive to the type of scattering (magnetic versus nonmagnetic). It is also noted that scaling with a T-dependent prefactor $\sigma_c(T)$ is not consistent with the simple single-parameter scaling hypothesis (Dobrosavljević et al., 1997), which allows only for a $d\sigma/dT < 0$ behavior in the conducting phase, but it does not violate any general principle and it is analogous to the MIT in 3D systems.

5.2.1.3 High-disorder samples

The most interesting situation is found in high-disorder samples, where electron–electron interactions are still strong (e.g., $r_s \approx 7$ near n_c). Indeed, the competition between disorder and interactions leads to the striking out-of-equilibrium or glassy behavior near the MIT and in the insulating regime. The manifestations of glassiness in *both* low- and high-disorder samples are discussed in Section 5.3. However, charge dynamics has been studied in more detail in high-disorder 2DESs, and there is ample evidence for the MIT and for the importance of long-range Coulomb interactions also in these systems (Popović, 2012). Thus one of the key questions that arises is, What is the nature of the MIT in a high-disorder 2DES with interactions? More precisely, is it dominated by disorder, or is it the same as the MIT in a low-disorder 2DES, which seems to be driven by interactions? The scaling behavior discussed below provides compelling evidence (a) for the existence of a metal–insulator QPT and (b) that sufficiently strong disorder changes the universality class of the MIT.

Detailed studies of transport and electron dynamics near the MIT have been performed on a set of Si-MOSFETs[d] with $\mu_{peak} \approx 0.06$ m^2/Vs. A typical $\sigma(n_s, T)$ is shown in Fig. 5.5 Because of the glassy fluctuations of σ with time t at low n_s and T, Fig. 5.5 actually shows the time-averaged conductivity $\langle\sigma\rangle$, and the error bars correspond to the size of the fluctuations with time. In this section, which focuses on the behavior of the average conductivity, the notation σ will be used instead of $\langle\sigma\rangle$ for simplicity. It is important to note that scaling of the average conductance is expected to work even in the presence of large fluctuations. For example, it has been demonstrated that for the case of Anderson transitions, scaling is, in fact, valid not only for the average conductance but also for all moments, even when relative fluctuations $\Delta\sigma/\langle\sigma\rangle \approx 1$ (Evers and Mirlin, 2008). Therefore, exploring the average conductance to get information about the MIT is well justified.

Similar to other high-disorder samples or samples with very low 4.2 K peak mobility ($\mu_{peak} < 0.1$ m^2/Vs), there is hardly any $d\sigma/dT < 0$ metallic behavior observed at high n_s. In analogy with studies of a low-disorder 2DES (Section 5.2.1.1), n_s^* is defined as the density where $d\sigma/dT$ changes sign and n_c is determined by extrapolating the activation energies in the insulating regime to zero. Surprisingly, here n_c turns out to be more than a factor of two smaller than n_s^* (Fig. 5.5). For $n_s > n_c$, the low-T data are best described by the metallic ($\sigma(T = 0) > 0$) power law $\sigma(n_s, T) = \sigma(n_s, T = 0) + b(n_s)T^{1.5}$ (Bogdanovich and Popović, 2002b). The surprising non-Fermi liquid (NFL) $T^{3/2}$ behavior is consistent with theory (Dalidovich and Dobrosavljević, 2002; Müller et al., 2012) for the transition region between a Fermi liquid and an (insulating) electron glass. Indeed (see Section 5.3.1), the transition into a charge (Coulomb) glass in high-disorder samples takes place as $T \rightarrow 0$ at a density n_g, such that $n_c < n_g < n_s^*$ (Fig. 5.5). The $T^{3/2}$ correction is characteristic of transport in the intermediate, $n_c < n_s < n_g$ region where the dynamics is glassy but where σ is still metallic [$\sigma(T \rightarrow 0) \neq 0$] albeit so small that $k_F l < 1$ (Fig. 5.6; k_F: Fermi wave

---

[d]The substrate bias $V_{sub} = -2$ V was applied to maximize μ_{peak} by removing the contribution of scattering by local magnetic moments (see Section 5.2.1.2 above), at least in the experimental T range.

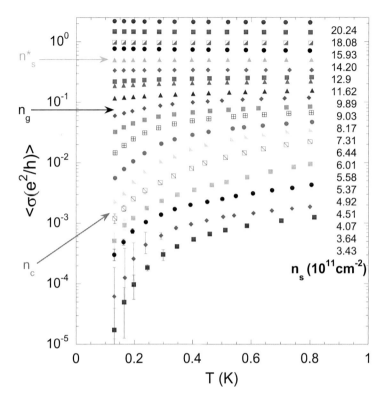

Figure 5.5 Time-averaged conductivity versus T for different n_s, as shown, in a high-disorder 2DES with $\mu_{\text{peak}} \approx 0.06$ m^2/Vs. The arrows mark three densities: $n_s^* = 12.9 \times 10^{11}$ cm^{-2} is the separatrix, $n_g = 7.5 \times 10^{11}$ cm^{-2} is the glass transition density, and $n_c = 5.0 \times 10^{11}$ cm^{-2} is the critical density for the MIT (the corresponding $r_s \approx 7$). Reprinted (figure) with permission from [Bogdanovich, S., and Popović, D. (2002b). *Phys. Rev. Lett.* **88**, p. 236401, erratum, *Phys. Rev. Lett.* **89**, 289904 (2002).] Copyright (2002) by the American Physical Society.

vector; l: mean free path). For $k_Fl < 1$, there is indeed no reason to expect standard Fermi liquid behavior. The simple form of $\sigma(T)$ in the intermediate, metallic phase allows a reliable extrapolation to $T = 0$. The extrapolated values of $\sigma(n_s, T = 0)$ go to zero at n_c that is in agreement with that obtained from the data on the insulating side of the MIT, and $\sigma_c(T) \propto T^x$ with $x = 1.5$. For clarity, the main experimental observations are summarized in Fig. 5.6.

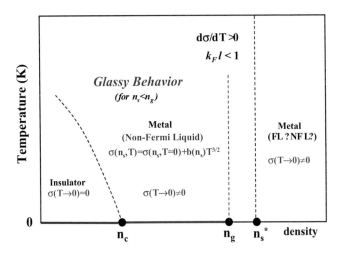

Figure 5.6 Experimental phase diagram of the 2DES in Si-MOSFETs. The MIT takes place at n_c and $T = 0$. The nature of the metallic phase at high n_s, such that $k_F l > 1$, is still under debate. The glass transition occurs at $T_g = 0$ for all $n_s < n_g$: the solid red line thus represents a line of quantum-critical points. In the intermediate, metallic glassy phase $n_c < n_s < n_g$, $\sigma(T)$ obeys a particular, non-Fermi liquid form. The aging properties of the glass change abruptly at n_c, indicating different natures of the insulating and metallic glass phases. In low-disorder samples, the intermediate phase vanishes: $n_c \lesssim n_s^* \approx n_g$.

Figure 5.7 demonstrates that near n_c, the data exhibit dynamical scaling $\sigma(n_s, T)/\sigma_c(T) \propto \sigma(n_s, T)/T^{1.5}$, a signature of the QPT, also in this system. The scaling parameter follows a power law, $T_0 \propto |\delta_n|^{z\nu}$ (Fig. 5.7 inset), in agreement with theoretical expectations near a QPT. The value of the critical exponents $z\nu \approx 2.1$ obtained here represents a major difference from the consistently lower $z\nu = 1.0$–1.7 found in a low-disorder 2DES (Feng et al., 1999, 2001; Kirkpatrick and Belitz, 2013; Kravchenko et al., 1995, 1996; Popović et al., 1997; Pudalov et al., 1998), indicating that sufficiently strong disorder changes the nature of the MIT from interaction driven in low-disorder samples to disorder dominated in high-disorder 2DES. Before discussing different universality classes further (Section 5.2.4), another central issue needs to be addressed first, namely the role of the range of electron–electron interactions in the 2D MIT.

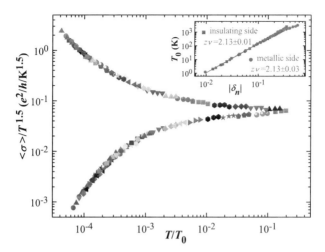

Figure 5.7 Scaling of $\sigma/\sigma_c \propto \sigma/T^x$, $x = 1.5$, for the high-disorder sample in Fig. 5.5 Different symbols correspond to n_s from 3.45×10^{11} cm^{-2} to 8.17×10^{11} cm^{-2}; $n_c = 5.22 \times 10^{11}$ cm^2. It was possible to scale the data below $\sim$0.3 K down to the lowest $T = 0.13$ K (the lowest $T/T_F \approx 0.003$). Inset: T_0 versus δ_n. The lines are fits with slopes $zv = 2.13 \pm 0.01$ and $zv = 2.13 \pm 0.03$ on the insulating and metallic sides, respectively. Reprinted (figure) with permission from [Lin, P. V., and Popović, D. (2015). *Phys. Rev. Lett.* **114**, p. 166401.] Copyright (2015) by the American Physical Society.

5.2.2 Effects of the Range of Electron–Electron Interactions

According to the scaling theory of localization (Abrahams et al., 1979), all electrons in a disordered, noninteracting 2D system become localized at $T = 0$. Therefore, interactions must play a key role in stabilizing a metallic ground state in two dimensions. An important question then is how the length scale of the Coulomb interactions controls the ground state and properties of the MIT. However, a vast majority of experiments on 2D systems (e.g., those discussed elsewhere in this chapter and in the chapter by A. A. Shashkin and S. V. Kravchenko) have been carried out on devices in which Coulomb interactions are *not* screened. At the same time, the use of a nearby metallic gate or ground plane to limit the range of the Coulomb interactions between charge carriers in 2D systems is a well-known technique that has been explored both theoretically (see, for example, Fregoso and Sá de Melo, 2013; Hallam et al., 1996; Ho et al., 2009; Peeters, 1984; Skinner and Fogler, 2010; Skinner and Shklovskii, 2010; Widom and Tao, 1988) and experimentally, for example, in the investigation of the melting of the Wigner crystal formed by electrons on a liquid He surface (Mistura et al., 1997).

In the context of the 2D MIT in low-disorder devices, screening by the gate has been used to explore the role of Coulomb interactions in the metallic (Ho et al., 2008) and insulator-like (Huang et al., 2014) regimes of a 2D hole system (2DHS) in AlGaAs/GaAs heterostructures and in the metallic regime of Si-MOSFETs (Tracy et al., 2009). The focus of those studies, however, was on the form of $\sigma(T)$, and especially on the sign of $d\sigma/dT$. In particular, there have been no studies of scaling and critical exponents, that is, of the properties of the quantum-critical point itself.

In the case of low-disorder 2DES with local magnetic moments, there have been no studies at all of the effects of screened Coulomb interactions.

On the other hand, in high-disorder Si-MOSFETs, studies of scaling behavior of $\sigma(n_s, T)$ on both metallic and insulating sides of the MIT were carried out on devices in which the long-range part of the Coulomb interaction is screened by the gate (Lin and Popović, 2015). The metallic gate at a distance d from the 2DES creates an image charge for each electron, modifying the Coulomb interaction from $\sim 1/r$ to $\sim [1/r - 1/\sqrt{r^2 + 4d^2}]$. When the mean carrier separation $a = (\pi n_s)^{1/2} \gg d$, this potential falls off in a dipole-like fashion, as $\sim 1/r^3$. Therefore, in Si-MOSFETs, the range of the electron–electron Coulomb interactions can be changed by varying the thickness of the oxide $d_{ox} = d$. Importantly, measurements were done on a set of MOSFETs that had been fabricated simultaneously as the high-disorder devices with long-range Coulomb interactions discussed in Section 5.2.1.3 (see also Figs. 5.5 and 5.7). In the latter case, $d_{ox} = 50$ nm, comparable to that in other Si-MOSFETs used in the vast majority of studies of the 2D MIT (Abrahams et al., 2001; Kravchenko and Sarachik, 2004; Popović, 2012; Spivak et al., 2010). In the $d_{ox} = 50$ nm samples, in the low-n_s regime of interest near the MIT, the corresponding $5.3 \lesssim d/a \leq 8.0$. On the other hand, in thin-oxide devices with $d_{ox} = 6.9$ nm (Lin and Popović, 2015), substantial screening by the gate is expected in the scaling regime of

n_s near the MIT, where $0.7 \lesssim d/a \lesssim 1.0$. For comparison, in ground-plane screening studies were carried out on low-disorder samples: $0.8 \leq d/a \leq 1.8$ (Tracy et al., 2009), $1.1 \lesssim d/a \leq 5$ (Huang et al., 2014), and $2 \leq d/a \leq 19$ (Ho et al., 2008).

In general, the results obtained on high-disorder, thin-oxide samples resemble those on thick-oxide devices, as follows.[e]

- At the lowest n_s, $\sigma(T)$ decreases exponentially with decreasing T, in agreement with the 2D variable-range hopping law, indicating an insulating ground state. The critical density n_c is determined from the vanishing of the activation energy, as described earlier.
- For $n_s > n_c$, the low-T data are best described by the metallic power law $\sigma(n_s, T) = \sigma(n_s, T = 0) + b(n_s)T^{1.5}$. This type of non-Fermi-liquid behavior was found to be a characteristic of the intermediate, metallic glassy phase observed in high-disorder samples with long-range Coulomb interactions (Section 5.2.1.3), as well as in a low-disorder 2DES in a parallel magnetic field (Section 5.2.3). At n_c, $\sigma_c \propto T^x$ with $x = 1.5$.
- On the metallic side of the MIT, the extrapolated $\sigma(n_s, T = 0)$ go to zero (Fig. 5.8b) at the same density as n_c obtained from the insulating side. The power-law behavior $\sigma(n_s, T = 0) \propto \delta_n^\mu$ (Fig. 5.8c) is in agreement with general expectations (Eq. 5.4). The critical exponent $\mu = 2.7 \pm 0.3$.
- Finally, in the vicinity of the MIT, the conductivity can be described by a scaling form (Eq. 5.2) (Fig. 5.8a); here $z\nu \approx 2.0$. As expected for a QPT, the same value of $z\nu$ is found, within experimental error, on both sides of the transition. The consistency of the scaling analysis is confirmed by comparing $\mu = x(z\nu) = 3.0 \pm 0.3$ obtained from scaling with $\mu = 2.7 \pm 0.3$ found from the $T = 0$ extrapolations of $\sigma(n_s, T = 0)$.

The most significant result is that the critical exponents in thin- and thick-oxide high-disorder devices are the same and thus not

---

[e]In analogy with Section 5.2.1.3, σ is used here to denote the time-averaged conductivity $\langle \sigma \rangle$.

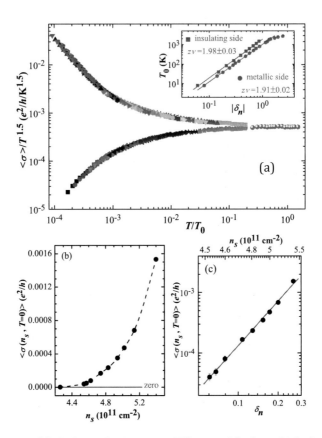

Figure 5.8 (a) Scaling of $\sigma/\sigma_c \propto \sigma/T^x$, $x = 1.5$, for a high-disorder sample with a thin oxide ($d_{ox} = 6.9$ nm), that is, with screened Coulomb interactions. Different symbols correspond to n_s from 3.40×10^{11} cm^{-2} to 6.70×10^{11} cm^{-2}; $n_c = 4.26 \times 10^{11}$ cm^2. It was possible to scale the data below about 1.5 K, over the range $0.007 \lesssim T/T_F \lesssim 0.04$. Inset: T_0 versus δ_n. The lines are fits with slopes $zv = 1.98 \pm 0.03$ and $zv = 1.91 \pm 0.02$ on the insulating and metallic sides, respectively. (b) $\sigma(n_s, T = 0)$ versus n_s. The dashed curve guides the eye. (c) $\sigma(n_s, T = 0)$ versus $\delta_n = (n_s - n_c)/n_c$, the distance from the MIT. The solid line is a fit with the slope equal to the critical exponent $\mu = 2.7 \pm 0.3$. Reprinted (figure) with permission from [Lin, P. V., and Popović, D. (2015). *Phys. Rev. Lett.* **114**, p. 166401.] Copyright (2015) by the American Physical Society.

sensitive to the range of the Coulomb interactions. Indeed, in such a disorder-dominated MIT, it is plausible that the length scale of the Coulomb interactions does not seem to play a major role. It is important to note, though, that there may be some other quantities that are more sensitive to the range of the Coulomb interactions and that would be affected by the proximity to the gate. For example, the fate of the glassy behavior in a 2DES with short-range Coulomb interactions remains an open question.

5.2.3 Effects of a Magnetic Field

Magnetic fields B applied parallel to the 2DES plane couple only to electrons' spins, and therefore, they are often used to probe the importance of spin, as opposed to charge, degrees of freedom. One of the main questions addressed in such studies in the context of the 2D MIT has been the fate of the metallic phase in a parallel B. High-disorder 2DESs have not been investigated yet, but on low-disorder samples, it has been established that the metallic phase and the MIT survive in high parallel B such that the 2DES is fully spin polarized. In particular, $(n_s, B, T = 0)$ phase diagrams have been determined both in the absence (Jaroszyński et al., 2004) and in the presence of local magnetic moments (Eng et al., 2002). Just like in $B = 0$, the properties of the MIT as a QPT have proved easier to study and analyze in systems with local moments, because of their stronger and better-defined $\sigma(T)$. In the absence of local moments, the situation is more subtle, as described below.

In a zero field, studies of charge dynamics in a low-disorder 2DES (with no local moments) have established (Section 5.3.2) that the onset of glassy behavior essentially coincides with the MIT, that is, $n_g \approx n_c$. A parallel B, however, gives rise to an intermediate, metallic glassy phase (Fig. 5.9) with the same, non-Fermi-liquid form $\sigma(n_s, B, T) = \sigma(n_s, B, T = 0) + b(n_s, B)T^{1.5}$ as what was observed in high-disorder samples at $B = 0$. Once again, such a simple and well-defined T dependence allows for reliable extrapolations to $T = 0$, finding $n_c(B)$, $\sigma_c \propto T^{1.5}$, and $\sigma(n_s, B, T = 0) \propto \delta_n^\mu$, $\mu \approx 1.5$, for a given B (Jaroszyński et al., 2004), in agreement with theoretical expectations near a QPT. Therefore, it is the emergence of the metallic glassy phase in $B \neq 0$ that makes it possible to

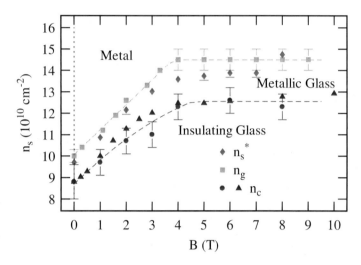

Figure 5.9 $T = 0$ phase diagram for a low-disorder 2DES in a parallel magnetic field. The dashed lines guide the eye. The n_c values are from Jaroszyński et al. (2004) (dots) and Shashkin et al. (2001) (triangles). The glass transition takes place at $n_g(B) > n_c(B)$, giving rise to an intermediate, metallic glass phase. The density at the separatrix $n_s^* \approx n_g$ within the error for all B. Reprinted (figure) with permission from [Jaroszyński, J., Popović, D., and Klapwijk, T. M. (2004). *Phys. Rev. Lett.* **92**, p. 226403.] Copyright (2004) by the American Physical Society.

determine n_c also from the *metallic* side of the MIT, the task that in the zero field, remains a challenge (Section 5.2.1.1). The remarkable agreement between $n_c(B)$ obtained from $\sigma(T)$ on both insulating and metallic sides of the MIT are strong evidence for the survival of the MIT and the metallic phase in parallel B. In contrast to highly disordered samples, here the intermediate phase spans a very narrow range of n_s, and therefore, it can be observed only if n_s is varied in fine steps. Finally, detailed studies of the behavior in the metallic glassy phase and the MIT have been performed so far only up to about 4 T in Fig. 5.9, the field above which the 2DES is fully spin polarized. Properties of the MIT at higher fields, that is, between a spin-polarized metal and a spin-polarized insulator, await future study.

In the presence of local magnetic moments, it is interesting that the $(n_s, B, T = 0)$ phase diagram (Eng et al., 2002) is quite similar to that shown in Fig. 5.9, even though the $d\sigma/dT$ behaviors in the metallic phase at $B = 0$ are strikingly different (Sections 5.2.1.1 and 5.2.1.2). This similarity probably results from the general expectation of a power-law shift of n_c with B in the case of a true MIT (Belitz and Kirkpatrick, 1994). Such a shift has been observed also in several 3D systems (Bogdanovich et al., 1997; Rosenbaum et al., 1989; Sarachik et al., 1998; Watanabe et al., 1999).

5.2.4 Possible Universality Classes of the 2D Metal–Insulator Transition

All the critical exponents have been summarized in Table 5.1 (Lin and Popović, 2015). On the basis of their values, it appears that the 2D MIT in Si-MOSFETs can be divided into three universality classes: (a) high disorder, (b) low disorder, and (c) low disorder in the presence of scattering by local magnetic moments. However, so far there have been no studies of high-disorder samples with local moments.

Table 5.1 also includes, where available, the values obtained in low parallel B (i.e., B is not high enough to fully spin polarize the 2DES Okamoto et al., 1999; Vitkalov et al., 2000). In low parallel B, $[n_c(B)/n_c(0) - 1] \propto B^\beta$ with $\beta = 1.0 \pm 0.1$ for low-disorder samples both in the absence (Dolgopolov et al., 1992; Jaroszyński et al., 2004; Sakr et al., 2001; Shashkin et al., 2001) and in the presence of scattering by local magnetic moments (Eng et al., 2002). It is apparent that such low fields do not seem to affect any of the critical exponents.

On the other hand, there is a major difference between the values of $z\nu$ in high- and low-disorder devices, indicating that sufficiently strong disorder changes the nature of the MIT from the interaction-driven to disorder-dominated. The possibility of a disorder-dominated 2D MIT has been demonstrated theoretically (Punnoose and Finkelstein, 2005) for both long-range and short-range interactions. Although there is currently no microscopic theory that describes the detailed properties of the observed disorder-dominated MIT, it is interesting that in the available theories (Belitz and Kirkpatrick, 1994; Punnoose and Finkelstein, 2005), the range of the Coulomb interactions does not play a significant role, consistent with experimental observations. In the

Table 5.1 Critical exponents x, $z\nu$, and calculated $\mu = x(z\nu)$ for 2D electron systems in Si-MOSFETs with different disorder [Lin and Popović, 2015]. "–" indicates that the data are either insufficient or unavailable. μ_{peak} is given in units of m^2/Vs, d_{ox} in nm, and n_c in 10^{11}cm^{-2}.

	High-disorder system		Special disorder: local magnetic moments		Low-disorder system	
	thin oxide	thick oxide				
μ_{peak}	0.04	0.06	~1		~1-3	
d_{ox}	6.9	50	43.5		40-600	
	$B=0$	$B=0$	$B=0$	$B\neq 0$	$B=0$	$B\neq 0$
n_c	4.2 ± 0.2	5.0 ± 0.3	0.5-1	$\{[n_c(B)/n_c(0)]-1\} \propto B$	~1	$\{[n_c(B)/n_c(0)]-1\} \propto B$
x	1.5 ± 0.1	1.5 ± 0.1	2.6 ± 0.4	2.7 ± 0.4	–	1.5 ± 0.1
$z\nu$	2.0 ± 0.1	2.1 ± 0.1	1.3 ± 0.1	0.9 ± 0.3	$1.0 - 1.7$	–
μ	2.7 ± 0.3	–	3.0 ± 0.1	3.0 ± 0.1	1-1.5	1.5 ± 0.1
$\mu = x(z\nu)$	3.0 ± 0.3	3.3 ± 0.4	3.4 ± 0.4	2.4 ± 1	–	–

interaction-driven MIT in low-disorder 2DESs (see also chapter by V. Dobrosavljević), the effect of the range of electron–electron interactions on the critical exponents still remains to be studied experimentally.

It should be also noted that percolation models (Stauffer and Aharony, 1994) cannot describe these findings. For example, the 2D percolation $\mu \simeq 1.3$, as opposed to the much larger experimental $\mu \simeq 3$ in high-disorder samples and low-disorder devices with local magnetic moments (Table 5.1). In fact, it is interesting that the same large $\mu \simeq 3$ is observed in those two types of samples, even though their values of $z\nu$ are very different ($z\nu \approx 2$ and $z\nu \approx 1.3$, respectively). Therefore, while $z\nu$ seems to depend on the amount of disorder, the exponent x instead appears to be more sensitive to the type of disorder (e.g., magnetic versus nonmagnetic).

To confirm the proposed universality classes of the 2D MIT (Table 5.1), which have been established on the basis of studies of Si-MOSFETs, it is clearly necessary to probe the behavior of 2D systems in other types of materials, in particular beyond conventional semiconductor heterostructures. New families of 2D crystals, formed by extracting atomically thin layers of materials with weak interlayer van der Waals interactions (Novoselov et al., 2005; Roldán et al., 2015), represent specially promising candidates for such investigations.

5.2.5 Metal–Insulator Transition in Novel 2D Materials

The 2D materials extracted from van der Waals solids, such as transition metal dichalcogenides (e.g., MoS_2, WS_2, $MoSe_2$, WSe_2), represent a new avenue for exploring quantum-critical phenomena and the effects of dimensionality on correlated electronic phases and may also lead to the development of new electronic and optoelectronic applications. Indeed, there is currently intensive activity in device fabrication based on 2D atomic layers, including FETs, photodetectors, light-emitting devices, etc. (Wang et al., 2012).

Recently, FETs with mobilities as high as ~ 0.03 m^2/Vs at a few Kelvin have been reported in MoS_2 (Radisavljevic and Kis, 2013) and ReS_2 (Pradhan et al., 2015). Those mobilities are comparable to μ_{peak} in high-disorder Si-MOSFETs (Section 5.2.1.3), which had

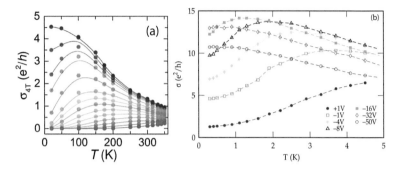

Figure 5.10 (a) Four-terminal conductivity of a few-layer ReS$_2$ versus T for different n_s, which are controlled by the substrate (back-gate) bias. Reprinted with permission from Pradhan, N. R., McCreary, A., Rhodes, D., Lu, Z., Feng, S., Manousakis, E., Smirnov, D., Namburu, R., Dubey, M., Walker, A. R. H., Terrones, H., Terrones, M., Dobrosavljevic, V., and Balicas, L. (2015). *Nano Lett.* **15**, 12, pp. 8377–8384. Copyright 2015 American Chemical Society. (b) σ versus T in a low-disorder Si-MOSFET for a fixed $n_s = 3.0 \times 10^{11}$ cm^{-2} and different V_{sub}, as shown. Reprinted (figure) with permission from [Feng, X. G., Popović, D., and Washburn, S. (1999). *Phys. Rev. Lett.* **83**, p. 368.] Copyright (1999) by the American Physical Society.

been fabricated using a commercial 0.25-µm Si technology (Taur and Ning, 1999). In both MoS$_2$ and ReS$_2$ FETs, metallic $d\sigma/dT < 0$ behavior was observed at high $n_s \gtrsim 10^{13}$ cm^{-2}, and the transition to $d\sigma/dT > 0$ temperature dependence at somewhat lower n_s was attributed to a 2D MIT. Although all the material parameters may not be known precisely (Pradhan et al., 2015), it is estimated that $r_s \approx 4$ near the apparent MIT in both types of devices, that is, a bit smaller than in high-disorder 2DES in Si where $r_s \approx 7$.

It is interesting that in ReS$_2$, $\sigma(T)$ exhibits nonmonotonic behavior at high n_s (Fig. 5.10a) reminiscent of a low-disorder 2DES in which scattering by local magnetic moments is controlled by varying the substrate bias. Figure 5.10b, for example, shows the effect of V_{sub} in a Si-MOSFET for a fixed $n_s > n_c$: while $d\sigma/dT < 0$ for a large negative $V_{sub} = -50$ V, reducing the negative V_{sub} leads to the emergence of a maximum in $\sigma(T)$, such that $d\sigma/dT > 0$ in the entire experimental T range for $V_{sub} = +1$ V. It is understood (Feng et al., 1999, 2001) that for T below the maximum, the scattering is dominated by local magnetic moments (Section 5.2.1.2; also

Fig. 5.3). Most notably, it has been shown (Pradhan et al., 2015) that $\sigma(T)$ of few-layer ReS$_2$ FETs at temperatures below the maximum obey dynamical scaling (Eq. 5.2) with the exponents $x \approx 2.3$, $zv \approx 1.3$, and $\mu \approx 2.9$, the latter being consistent with $\mu = x(zv) \approx 3.0$. It is most striking that these values indeed agree, within error, with the exponents established for (low-disorder) samples in which local moments dominate (Table 5.1).[f]

Although the agreement between critical exponents obtained on ReS$_2$ and those on 2DES in Si is encouraging, there are several caveats. For example, in ReS$_2$ there is some uncertainty in the determination of n_s as a function of V_{sub} so that depending on the method used, the data analysis may yield a different set of exponents (e.g., $x \approx 3.4$, $zv \approx 0.6$, $\mu \approx 2.1$; Pradhan et al., 2015). In addition, scaling was performed at fairly high $T \lesssim T_F$ and over a limited range of n_s in the insulating regime. Therefore, not only do the measurements need to be extended to much lower T and n_s, but also it is important to gain a better understanding of the basic FET and material characteristics before reliable results on the 2D MIT can be obtained on ReS$_2$ and other novel 2D materials.

5.3 Charge Dynamics Near the 2D Metal–Insulator Transition and the Nature of the Insulating State

Section 5.2 focused on the critical region near the 2D MIT, in particular on describing conductivity measurements that have demonstrated dynamical scaling (Eq. 5.2), the main signature of the MIT as a QPT. The critical exponents, which have been determined for both interaction-driven and disorder-dominated MIT (Section 5.2.4), represent a property of the quantum-critical point. However, these studies do not provide information about the nature of the metallic and insulating phases. The chapter by A. A. Shashkin and S. V. Kravchenko discusses various experimental results obtained in the metallic regime of low-disorder 2DESs. In contrast, this section

---

[f]As noted in Section 5.2.4, there have been no studies of high-disorder samples with local magnetic moments.

focuses on experiments that probe the nature of the *insulating* state, as well as charge dynamics *across the MIT*, in both high- and low-disorder 2DESs.

There is substantial evidence that in many materials near the MIT, both strong electronic correlations and disorder play an important role, and thus their competition is expected to lead to glassy behavior of electrons, in analogy with other frustrated systems (Miranda and Dobrosavljević, 2005, 2012). A common denominator for all glasses is the existence of a complex or "rugged" free-energy landscape, consisting of a large number of metastable states, separated by barriers of different heights. This results in phenomena such as slow, nonexponential relaxations, divergence of the equilibration time, and breaking of ergodicity, that is, the inability of the system to equilibrate on experimental timescales. Therefore, such out-of-equilibrium systems also exhibit aging effects (Rubi and Perez-Vicente, 1997; Struik, 1978), where the response to an external excitation (i.e., relaxation) depends on the system history in addition to the time t. A detailed analysis of temporal fluctuations (noise) of the relevant observables yields complementary information on configurational rearrangements or transitions between metastable states.[g] Non-Gaussian distributions of various observables in glassy systems have been reported (Berthier et al., 2011), reflecting the presence of large, *collective* rearrangements. Therefore, the two basic ways to probe the dynamics of glassy systems involve studies of relaxations and fluctuations.

Most experimental studies of charge or Coulomb glasses have focused on situations where electrons are strongly localized due to disorder, that is, deep in the insulating regime and far from the MIT (Amir et al., 2011). In recent years, however, studies of both relaxations and fluctuations (or noise) in Si-MOSFETs have provided evidence for out-of-equilibrium or glassy dynamics of the 2DES in the insulating regime, near the MIT, and just on the metallic side of the transition in the intermediate, metallic glassy phase (Fig. 5.6).

---

[g]In equilibrium systems, the connection between spontaneous fluctuations of a variable and the response of such a variable to a small perturbation in its conjugated field is given by the fluctuation–dissipation relation. See (Leuzzi and Nieuwenhuizen, 2008) for the review and discussion of thermodynamics of out-of-equilibrium systems.

Those results, described below (see also Popović, 2012), impose strong constraints on the theories for the 2D MIT and should be also helpful in understanding the complex behavior near the MIT in a variety of strongly correlated materials.

5.3.1 High-Disorder 2D Electron Systems

Glassy charge dynamics in a high-disorder 2DES was probed using several different experimental protocols, all of which can be divided into two groups: One of them involves applying a large (with respect to E_F) perturbation to the system and studying the relaxations of conductivity, and the other one involves a study of conductivity fluctuations with time as a result of a small perturbation. In all experiments, $k_B T$ was the lowest energy scale.

A *large perturbation* was applied to the 2DES by making a large charge in V_g or carrier density, that is, such that $k_B T \ll E_F < \Delta E_F$. In one protocol, which involved a study of the relaxations $\sigma(t)$ following a rapid change of n_s, the following key manifestations of glassiness were established (Jaroszyński and Popović, 2006) for all n_s below the glass transition density n_g ($n_c < n_g$, that is, n_g is on the metallic side of the MIT; Fig. 5.6).

- The temperature dependence of the equilibrium time τ_{eq} obeys a simply activated form, so $\tau_{eq} \to \infty$ as $T \to 0$. The diverging equilibrium time means that, strictly speaking, the system cannot reach equilibrium only at $T = 0$, that is, the glass transition temperature $T_g = 0$.
- At low enough T, however, τ_{eq} can easily exceed experimental times (e.g., at 1 K, τ_{eq} is estimated to exceed the age of the universe by several orders of magnitude!), so the system appears glassy: For $t < \tau_{eq}$, the relaxations obey a nonexponential form, which reflects the existence of a broad distribution of relaxation times.
- Nonexponential relaxations obey dynamical scaling $\sigma(t, T)/\sigma_0(T) \propto t^{-\alpha(n_s)} \exp[-(t/\tau(n_s, T))^{\beta(n_s)}]$, where $0 < \alpha(n_s) < 0.4, 0.2 < \beta(n_s) < 0.45$, such that in the $T \to 0$ limit, the relaxations attain a pure power-law form $\sigma/\sigma_0 \propto t^{-\alpha}$. The dynamical scaling and the power-law relaxation

at T_g are consistent with the general scaling arguments (Hohenberg and Halperin, 1977) near a continuous phase transition occurring at $T_g = 0$, similar to the discussion in Section 5.1.

A key characteristic of relaxing glassy systems is the loss of time translation invariance, reflected in aging effects (Hodge, 1995; Rubi and Perez-Vicente, 1997; Struik, 1978). Therefore, in another protocol, relaxations $\sigma(t)$ were studied after a temporary change of n_s during the waiting time t_w. The sample history was varied by changing t_w and T for several initial (final) n_s. The main results include the following.

- It was demonstrated that the 2DES exhibits aging, and the conditions that lead to memory loss and nonmonotonic response were identified precisely (Jaroszyński and Popović, 2007b).
- There is an abrupt change in the nature of the glassy phase exactly at the 2D MIT itself, before glassiness disappears completely at a higher density n_g: (a) While the so-called full aging[h] is observed in the insulating regime ($n_s < n_c$), there are significant departures from full aging in the metallic glassy phase, that is, for $n_c < n_s < n_g$; (b) the amplitude of the relaxations peaks just below the MIT, and it is strongly suppressed in the insulating phase (Jaroszyński and Popović, 2007a, 2009).
- As the system ages and slowly approaches equilibrium, the non-Gaussian conductance noise becomes increasingly Gaussian (Lin et al., 2012), similar to the behavior of a great variety of out-of-equilibrium systems.

The results of the aging studies represent strong evidence that the insulating glassy phase and the metallic glassy phase are different. It should be also noted that in the mean-field models of glasses, for example, two different cases are distinguished: one where full aging is expected and the other where no t/t_w scaling is expected (Bouchaud et al., 1997). Therefore, the difference in the

---

[h]In the case of full or simple aging, the aging function $\sigma(t, t_w)$ exhibits scaling with t/t_w.

aging properties below and above n_c puts constraints on the theories of glassy freezing and its role in the physics of the 2D MIT.

In the second group of experiments, a *small perturbation* was applied either by making a small change in n_s, such that $k_B T < \Delta E_F \ll E_F$ (Bogdanović and Popović, 2002a, 2002b; Jaroszyński et al. 2002; Lin et al. 2012), or by cooling, that is, making a small change in T, such that $k_B T < k_B \Delta T \ll E_F$ (Lin et al., 2012). In both cases, there were no observable relaxations of σ, but large non-Gaussian conductance noise emerged for $n_s < n_g$. The non-Gaussian nature of the noise indicates that the fluctuating units are correlated.

The noise was studied by analyzing the full probability distribution of the fluctuations (or the probability density function [PDF]), power spectrum, and the so-called second spectrum, which is a fourth-order noise statistic. The analysis indicates that the slow, correlated behavior is consistent with the so-called hierarchical picture of glassy dynamics, similar to conventional, metallic spin glasses (Binder and Young, 1986). In that scenario, the system wanders collectively between many metastable states related by a kinetic hierarchy. Metastable states correspond to the local minima, or valleys, in the free-energy landscape, separated by barriers with a wide, hierarchical distribution of heights and, thus, relaxation times. Intervalley transitions, which are reconfigurations of a large number of electrons, thus lead to the observed strong, correlated, $1/f$-type noise.

Therefore, the 2DES has many characteristics in common with a large class of both 2D and 3D out-of-equilibrium systems, strongly suggesting that many such universal features are robust manifestations of glassiness, regardless of the dimensionality of the system. However, it appears that there are also some effects that may be unique to Coulomb glasses. This is illustrated in Fig. 5.11, for example, where PDFs obtained after cooling (left column) are compared to those obtained after a subsequent large change of n_s during the waiting time t_w (right column). What is striking is that even though the PDFs were measured under exactly the same experimental conditions, they look remarkably different for all $n_s < n_g$. In particular, the non-Gaussian PDFs obtained after cooling are smooth, single-peaked functions, reminiscent of PDFs in a variety of systems displaying critical (Bramwell et al., 1998; Joubaud et al., 2008),

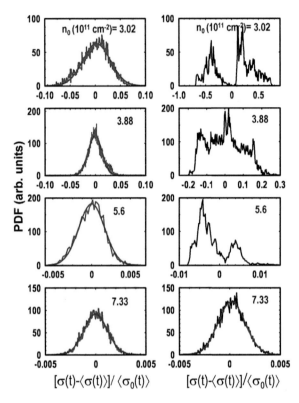

Figure 5.11 History dependence of the PDFs of the fluctuations measured in a high-disorder 2DES at $T = 0.24$ K for several carrier densities $n_0(10^{11}$ cm^{-2}), as shown; $n_g \approx 7.5 \times 10^{11}$ cm^{-2}. $\langle\sigma_0\rangle$ is the time-averaged conductivity corresponding to n_0 at the measurement T, obtained after cooling. The red curves are fits to a Gaussian distribution. Left column: PDFs of the noise after cooling from 10 K to 0.24 K for each given n_0. The cooling time was at least one hour long. In this protocol, $\langle\sigma(t)\rangle \equiv \langle\sigma_0\rangle$, that is, there are no observable relaxations after cooling. Right column: PDFs of the noise measured after a subsequent change of n_s from n_0 to a much higher value $n_1 = 20.26 \times 10^{11}$ cm^{-2} during $t_w = 1000$ s. In this protocol, $\langle\sigma(t)\rangle$ describes the slowly relaxing background. Reprinted (figure) with permission from [Lin, P. V., Shi, X., Jaroszynski, J., and Popović, D. (2012). *Phys. Rev. B* **86**, p. 155135.] Copyright (2012) by the American Physical Society.

glassy (Berthier et al., 2011), or other out-of-equilibrium behavior (e.g., the Danube water level Bramwell et al., 2002). In all these systems, the PDFs are skewed, resembling a zero-centered Gaussian, which describes pseudoequilibrium fluctuations, with one (exponential) tail that is due to large, rare events. In contrast, a temporary change in n_s results in complex, multipeaked, random-looking PDFs. Similar complicated, multipeaked PDFs were observed also after a small change of n_s (not shown). Therefore, these results demonstrate not only that noise depends on history, as may be expected in a glassy system, but also that the change of n_s has a qualitatively different and more dramatic effect than ΔT. In fact, the results strongly suggest that the density change reshuffles all energies, because of the Coulomb interactions, thus modifying the free-energy landscape of the 2DES. For this reason, theoretical modeling of the glassy dynamics in this system might be considerably more difficult than in some other types of glassy materials.

5.3.2 Low-Disorder 2D Electron Systems

Studies of charge dynamics in low-disorder 2DESs have so far included only small-perturbation protocols, that is, measurements of the conductance fluctuations following a small change of n_s, including the analysis of the higher-order statistics (Jaroszyński et al., 2002). Qualitatively, the same behavior was observed as in high-disorder samples: there is a well-defined density n_g below which the noise becomes non-Gaussian and increases by several orders of magnitude as n_s or T are reduced. This slow, correlated noise is consistent with the hierarchical pictures of glassy dynamics, as described above. The only difference is that, in a low-disorder 2DES, the intermediate glassy phase practically vanishes and the glass transition coincides with the MIT ($n_g \approx n_c$).

Noise measurements were performed also in parallel B (Jaroszyński et al., 2004). By adopting the same criteria for the glass transition as in zero field, it was possible to determine $n_g(B)$ shown in Fig. 5.9, identify the emergence of the intermediate, metallic glassy phase (see Section 5.2.3), and establish that charge, not spin, degrees of freedom are responsible for glassy ordering. Therefore, the results demonstrate that the 2D MIT is closely related to the melting of this Coulomb glass.

In fact, experiments on both high- and low-disorder 2DESs in Si strongly support theoretical proposals describing the 2D MIT as the melting of a Coulomb glass (Chakravarty et al., 1999; Dalidovich and Dobrosavljević, 2002; Dobrosavljević et al., 2003; Pastor and Dobrosavljević, 1999; Thakur and Neilson, 1996, 1999). In particular, a model with Coulomb interactions and sufficient disorder (Dobrosavljević et al., 2003) predicts the emergence of an intermediate metallic glass phase. The theoretical work, however, still needs to be extended to studies of the critical behavior, including critical exponents. Experimentally, it would be interesting to perform relaxation studies on low-disorder samples to look for aging phenomena, which may also provide further insights into the nature of the insulating, Coulomb glass phase.

5.4 Conductor–Insulator Transition and Charge Dynamics in Quasi-2D Strongly Correlated Systems

Many novel materials are created by doping an insulating host and thus are close to a conductor–insulator transition. For example, there is now broad agreement (Lee et al., 2006) that the problem of high-temperature superconductivity in copper oxides is synonymous to that of doping of a Mott insulator. Arguments have been put forward that in weakly doped Mott insulators near the MIT, the system will settle for a nanoscale phase separation between a conductor and an insulator (Dagotto, 2002; Gor'kov and Sokol, 1987; Kivelson et al., 2003). This leads to the possibility for a myriad of competing charge configurations and the emergence of the associated glassy dynamics, perhaps even in the absence of disorder (Schmalian and Wolynes, 2000). Indeed, cuprates and many other materials exhibit various complex phenomena due to the existence of several competing ground states (Dagotto, 2005), thus providing strong impetus toward a better understanding of the MIT and the behavior of the charge degrees of freedom. However, experimental studies near the conductor–insulator transition in many materials, such as cuprates, are complicated by the accompanying changes in magnetic or structural symmetry. On the other hand, those materials are usually layered, with weak interlayer coupling, so in most instances, they behave effectively as 2D systems. Therefore, comparative studies of the MIT and charge dynamics in 2DESs in semiconductor heterostructures and in strongly correlated quasi-2D materials, such as cuprates, present an especially promising approach in addressing the problem of complexity near the MIT in strongly correlated systems. Such studies should make it possible to separate out effects that are more universal from those that are material specific.

In $La_{2-x}Sr_xCuO_4$ (LSCO), the prototypical cuprate high-temperature superconductor, conductance (or resistance) noise spectroscopy was employed at very low T (Raičević et al., 2008, 2011) to probe charge dynamics in the lightly doped regime where the ground state is insulating. Here the charge carriers, doped holes, seem to populate areas that separate the hole-poor antiferromagnetic (AF) domains located in CuO_2 planes. The magnetic moments in different domains are known to undergo cooperative freezing at a temperature T_{SG} (approx. a few K) into a *cluster spin-glass* phase. The noise measurements demonstrated the emergence of slow, correlated dynamics and nonergodic behavior at very low $T \ll T_{SG}$, deep inside the spin-glass phase, which rules out spins as the origin of the observed glassy dynamics. In addition, all the noise characteristics were found to be insensitive to both the magnetic field and the magnetic history, further indicating that the observed glassiness reflects the dynamics of charge, not spins. This is analogous to the magnetic insensitivity of the noise in a glassy, fully spin-polarized 2DES in Si (Jaroszyński et al., 2004). The gradual enhancement of the glassy behavior in LSCO with decreasing T strongly suggests that the phase transition to a charge-glass state occurs at $T_g = 0$, similar to a 2DES (Section 5.3). In contrast to a 2DES, however, the non-Gaussian statistics in LSCO is not consistent with the hierarchical picture of glasses but, rather, reflects the presence of some characteristic length scale. This result supports the picture of spatial segregation of holes into interacting, hole-rich droplets or clusters, which are separated by hole-poor AF domains.

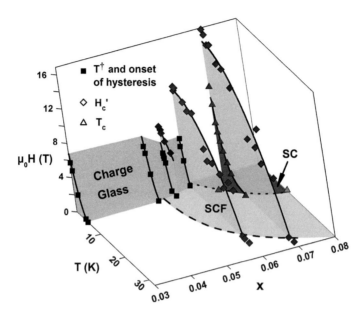

Figure 5.12 Phase diagram shows the evolution of the glassy region and the emergence of superconducting fluctuations (SCFs) and superconductivity (SC) with doping, temperature, and magnetic field in $La_{2-x}Sr_xCuO_4$ (Shi et al., 2012). The extent of the glassy regime does not depend on the field orientation. The range of SCFs is shown for the field applied perpendicular to CuO_2 planes. Solid and dashed lines guide the eye. Different colors of symbols for $H'_c(T)$ and $T_c(H)$ correspond to different values of doping.

Additional evidence for charge-glass behavior at low T in lightly doped, insulating LSCO was found from dielectric (Jelbert et al., 2008) and magnetotransport (Raičević et al., 2010) studies. The observed history-dependent resistance and hysteretic magnetoresistance (Raičević et al., 2008, 2010, 2012) were then used to explore the key question in the physics of cuprates, namely how such an insulating, dynamically heterogeneous ground state evolves with doping and gives way to high-temperature superconductivity (Shi et al., 2012). The (x, T, H) phase diagram (Fig. 5.12), where x (in this Section) is doping, H is a magnetic field perpendicular to CuO_2 planes, shows that a collective, glassy state of charge clusters located in CuO_2 planes is suppressed by increasing the doping. At the same time, adding charge carriers leads to the formation of localized

Cooper pairs (or superconducting fluctuations) already within this insulating, intrinsically heterogeneous charge-ordered state, consistent with the so-called Bose glass picture for the superconductor–insulator transition (SIT) (Fisher et al., 1990). Surprisingly, it was also found (not shown) that the superconducting fluctuations on the insulating side were quenched at low temperatures by the charge-glass order. Therefore, the pair localization and the onset of SIT in LSCO are influenced by a competing charge order, and not merely by disorder, as seems to be the case in some conventional superconductors. Those observations provide a new perspective on the mechanism for the SIT.

The experiments discussed above were carried out on samples in which the number of carriers was varied by chemical doping. To study scaling associated with the $T = 0$ SIT, however, that method has the disadvantage that the carrier concentration cannot be tuned continuously, and also it alters the level of disorder in the material. Therefore, electrostatic charging, similar to that in semiconductor heterostructures, would be preferable. Recently, it has become possible to electrostatically induce large concentration changes in a variety of novel materials (Ahn et al., 2006). In cuprates, ionic liquids have been used to make FET-like devices to study scaling near the SIT in thin films of hole-doped $La_{2-x}Sr_xCuO_4$ (Bollinger et al., 2011), $YBa_2Cu_3O_{7-x}$ (Leng et al., 2011) and $La_2CuO_{4+\delta}$ (Garcia-Barriocanal et al., 2013), and electron-doped $Pr_{2-x}Ce_xCuO_4$ (Zeng et al., 2015). The observed scaling appeared consistent with the so-called bosonic picture (Fisher et al., 1990), in which Cooper pairs are localized on the insulating side of the SIT and the critical resistivity is independent of T.[i] The obtained critical exponents $z\nu$ were 1.5, 2.2, 1.2, and 2.4, respectively, but the reasons for that difference are not understood. Moreover, in contrast to the 2DES in Si (Section 5.2), the scaling was done over a limited range of parameters: typically, T/T_0 spans only about an order of magnitude, and the lowest T is approx. a few Kelvin. Clearly, it would be important to extend these studies to much lower T. Interestingly, it was found that scaling in $YBa_2Cu_3O_{7-x}$ breaks down below 6 K, suggesting that the SIT may

---

[i]The additional presence of some fermionic excitations at finite T on the insulating side was suggested for the electron-doped cuprate (Zeng et al., 2015).

not be direct and may involve an intermediate phase (Leng et al., 2011). Unfortunately, there were not sufficient data available to resolve this issue (the lowest T in that experiment was 2 K), so that the questions about the nature of the carrier-concentration-driven SIT and whether it involves an intermediate phase in the $T \to 0$ limit remain open for all cuprates.

Finally, in analogy to a 2DES in Si (Section 5.3), electrostatic doping opens up an exciting possibility to investigate glassy relaxations that may accompany QPTs in cuprates and other strongly correlated systems. Such studies should provide a much better understanding of the phases and quantum criticality.

5.5 Conclusions

Experimental studies of the critical behavior of conductivity in a variety of 2DESs in Si provide strong evidence for the existence of a sharp, $T = 0$ MIT at low electron densities regardless of the amount of disorder. The critical exponents obtained from dynamical scaling suggest that there are several universality classes of the 2D MIT, depending on the amount and type of disorder. Since other types of experiments carried out on the metallic side of the transition indicate that the MIT in low-disorder samples is driven by electron–electron interactions, this implies that the MIT in high-disorder samples is dominated by disorder. In both cases, however, Coulomb interactions between electrons must play a key role in stabilizing the metallic phase.

The role of long-range Coulomb interactions is further revealed in the studies of charge dynamics across the MIT, which demonstrate that in both low- and high-disorder systems, the insulating state is a Coulomb glass. The peculiarity of a 2DES with a high amount of disorder is the emergence of an intermediate phase between the metal and the insulator, which is poorly metallic and glassy. The aging properties demonstrate, however, that the nature of the metallic glassy phase is different from that of the insulating Coulomb glass.

While the above experimental findings await theoretical description and understanding, novel 2D materials, such as those extracted from van der Waals solids, as well as quasi-2D strongly correlated materials, including cuprates, present promising new avenues for testing the generality of the observed phenomena and gaining further insight into the problem of the MIT.

Acknowledgments

This work was supported by NSF grant no. DMR-1307075 and the National High Magnetic Field Laboratory through NSF Cooperative agreement no. DMR-1157490 and the State of Florida.

References

Abrahams, E., Anderson, P. W., Licciardello, D. C., and Ramakrishnan, T. V. (1979). *Phys. Rev. Lett.* **42**, pp. 673–676.

Abrahams, E., Kravchenko, S. V., and Sarachik, M. P. (2001). *Rev. Mod. Phys.* **73**, pp. 251–266.

Ahn, C. H., Bhattacharya, A., Di Ventra, M., Eckstein, J. N., Frisbie, C. D., Gershenson, M. E., Goldman, A. M., Inoue, I. H., Mannhart, J., Millis, A. J., Morpurgo, A. F., Natelson, D., and Triscone, J.-M. (2006). *Rev. Mod. Phys.* **78**, pp. 1185–1212.

Altshuler, B. L., Maslov, D. L., and Pudalov, V. M. (2001). *Physica (Amsterdam)* **9E**, p. 209.

Amir, A., Oreg, Y., and Imry, Y. (2011). *Annu. Rev. Condens. Matter Phys.* **2**, p. 235.

Amit, D. J., and Peliti, L. (1982). *Ann. Phys.* **140**, pp. 207–231.

Ando, T., Fowler, A. B., and Stern, F. (1982). *Rev. Mod. Phys.* **54**, p. 437.

Belitz, D., and Kirkpatrick, T. R. (1994). *Rev. Mod. Phys.* **66**, p. 261.

Belitz, D., and Kirkpatrick, T. R. (1995). *Z. Phys. B* **98**, p. 513.

Berthier, L., Biroli, G., Bouchaud, J.-P., Cipelletti, L., and van Saarloos, W. (eds.) (2011). *Dynamic Heterogeneities in Glasses, Colloids and Granular Media* (Oxford University Press, UK).

Binder, K., and Young, A. P. (1986). *Rev. Mod. Phys.* **58**, p. 801.

Bogdanovich, S., Dai, P., Sarachik, M. P., Dobrosavljević, V., and Kotliar, G. (1997). *Phys. Rev. B* **55**, p. 4215.

Bogdanovich, S., and Popović, D. (2002a). *Physica E* **12**, p. 604.

Bogdanovich, S., and Popović, D. (2002b). *Phys. Rev. Lett.* **88**, p. 236401, erratum, *Phys. Rev. Lett.* **89**, 289904 (2002).

Bollinger, A. T., Dubuis, G., Misewich, J. Y. D. P. J., and Božović, I. (2011). *Nature* **472**, pp. 458–460.

Bouchaud, J.-P., Cugliandolo, L. F., Kurchan, J., and Mezard, M. (1997). Out of equilibrium dynamics in spin-glasses and other glassy systems, in *Spin Glasses and Random Fields*, ed. A. P. Young (World Scientific, Singapore).

Bramwell, S. T., Fennell, T., Holdsworth, P. C. W., and Portelli, B. (2002). *Europhys. Lett.* **57**, 3, p. 310.

Bramwell, S. T., Holdsworth, P. C. W., and Pinton, J.-F. (1998). *Nature* **396**, pp. 552–554.

Castellani, C., Kotliar, G., and Lee, P. A. (1987). *Phys. Rev. Lett.* **59**, p. 323.

Chakravarty, S., Kivelson, S., Nayak, C., and Voelker, K. (1999). *Philos. Mag. B* **79**, p. 859.

Dagotto, E. (2002). *Nanoscale Phase Separation and Colossal Magnetoresistance* (Springer-Verlag, Berlin).

Dagotto, E. (2005). *Science* **309**, pp. 257–262.

Dalidovich, D., and Dobrosavljević, V. (2002). *Phys. Rev. B* **66**, p. 081107.

Dobrosavljević, V. (2012). Introduction to metal-insulator transitions, in *Conductor-Insulator Quantum Phase Transitions*, pp. 3–63, eds. V. Dobrosavljević, N. Trivedi, and J. M. Valles, Jr. (Oxford University Press).

Dobrosavljević, V., Abrahams, E., Miranda, E., and Chakravarty, S. (1997). *Phys. Rev. Lett.* **79**, pp. 455–458.

Dobrosavljević, V., Tanasković, D., and Pastor, A. A. (2003). *Phys. Rev. Lett.* **90**, p. 016402.

Dolgopolov, V. T., Kravchenko, G. V., Shashkin, A. A., and Kravchenko, S. V. (1992). *JETP Lett.* **55**, pp. 701–705.

Eng, K., Feng, X. G., Popović, D., and Washburn, S. (2002). *Phys. Rev. Lett.* **88**, p. 136402.

Evers, F., and Mirlin, A. D. (2008). *Rev. Mod. Phys.* **80**, pp. 1355–1417.

Feng, X. G., Popović, D., and Washburn, S. (1999). *Phys. Rev. Lett.* **83**, p. 368.

Feng, X. G., Popović, D., Washburn, S., and Dobrosavljević, V. (2001). *Phys. Rev. Lett.* **86**, p. 2625.

Fisher, M. P. A., Grinstein, G., and Girvin, S. M. (1990). *Phys. Rev. Lett.* **64**, pp. 587–590.

Fletcher, R., Pudalov, V. M., Radcliffe, A. D. B., and Possanzini, C. (2001). *Semicond. Sci. Tech.* **16**, p. 386.

Fregoso, B. M., and Sá de Melo, C. A. R. (2013). *Phys. Rev. B* **87**, p. 125109.

Garcia-Barriocanal, J., Kobrinskii, A., Leng, X., Kinney, J., Yang, B., Snyder, S., and Goldman, A. M. (2013). *Phys. Rev. B* **87**, p. 024509.

Goldenfeld, N. (1992). *Lectures on Phase Transitions and the Renormalization Group* (Addison-Wesley).

Gor'kov, L. P., and Sokol, A. V. (1987). *JETP Lett.* **46**, p. 420.

Hallam, L. D., Weis, J., and Maksym, P. A. (1996). *Phys. Rev. B* **53**, pp. 1452–1462.

Ho, L. H., Clarke, W. R., Micolich, A. P., Danneau, R., Klochan, O., Simmons, M. Y., Hamilton, A. R., Pepper, M., and Ritchie, D. A. (2008). *Phys. Rev. B* **77**, p. 201402.

Ho, L. H., Micolich, A. P., Hamilton, A. R., and Sushkov, O. P. (2009). *Phys. Rev. B* **80**, p. 155412.

Hodge, I. M. (1995). *Science* **267**, p. 1945.

Hohenberg, P. C., and Halperin, B. I. (1977). *Rev. Mod. Phys.* **49**, p. 435.

Huang, J., Pfeiffer, L. N., and West, K. W. (2014). *Phys. Rev. Lett.* **112**, p. 036803.

Jaroszyński, J., and Popović, D. (2006). *Phys. Rev. Lett.* **96**, p. 037403.

Jaroszyński, J., and Popović, D. (2007a). *Phys. Rev. Lett.* **99**, p. 216401.

Jaroszyński, J., and Popović, D. (2007b). *Phys. Rev. Lett.* **99**, p. 046405.

Jaroszyński, J., and Popović, D. (2009). *Physica B* **404**, p. 466.

Jaroszyński, J., Popović, D., and Klapwijk, T. M. (2002). *Phys. Rev. Lett.* **89**, p. 276401.

Jaroszyński, J., Popović, D., and Klapwijk, T. M. (2004). *Phys. Rev. Lett.* **92**, p. 226403.

Jelbert, G. R., Sasagawa, T., Fletcher, J. D., Park, T., Thompson, J. D. and Panagopoulos, C. (2008). *Phys. Rev. B* **78**, p. 132513.

Joubaud, S., Petrosyan, A., Ciliberto, S., and Garnier, N. B. (2008). *Phys. Rev. Lett.* **100**, p. 180601.

Kirkpatrick, T. R., and Belitz, D. (1994). *Phys. Rev. Lett.* **74**, p. 1178.

Kirkpatrick, T. R., and Belitz, D. (2013). *Phys. Rev. Lett.* **110**, p. 035702.

Kivelson, S. A., Bindloss, I. P., Fradkin, E., Oganesyan, V., Tranquada, J. M., Kapitulnik, A., and Howald, C. (2003). *Rev. Mod. Phys.* **75**, pp. 1201–1241.

Kravchenko, S. V., Mason, W. E., Bowker, G. E., Furneaux, J. E., Pudalov, V. M. and D'Iorio, M. (1995). *Phys. Rev. B* **51**, p. 7038.

Kravchenko, S. V., and Sarachik, M. P. (2004). *Rep. Prog. Phys.* **67**, p. 1.

Kravchenko, S. V., Simonian, D., Sarachik, M. P., Mason, W., and Furneaux, J. E. (1996). *Phys. Rev. Lett.* **77**, pp. 4938–4941.

Lee, P. A., Nagaosa, N., and Wen, X.-G. (2006). *Rev. Mod. Phys.* **78**, pp. 17–85.

Lee, P. A., and Ramakrishnan, T. V. (1985). *Rev. Mod. Phys.* **57**, p. 287.

Leng, X., Garcia-Barriocanal, J., Bose, S., Lee, Y., and Goldman, A. M. (2011). *Phys. Rev. Lett.* **107**, p. 027001.

Leuzzi, L., and Nieuwenhuizen, T. M. (2008). *Thermodynamics of the Glassy State* (Taylor & Francis, New York).

Lin, P. V., and Popović, D. (2015). *Phys. Rev. Lett.* **114**, p. 166401.

Lin, P. V., Shi, X., Jaroszynski, J., and Popović, D. (2012). *Phys. Rev. B* **86**, p. 155135.

Miranda, E., and Dobrosavljević, V. (2005). *Rep. Prog. Phys.* **68**, p. 2337.

Miranda, E., and Dobrosavljević, V. (2012). Dynamical mean-field theories of correlation and disorder, pp. 161–243, in *Conductor-Insulator Quantum Phase Transitions*, eds. V. Dobrosavljević, N. Trivedi, and J. M. Valles, Jr. (Oxford University Press).

Mistura, G., Günzler, T., Neser, S., and Leiderer, P. (1997). *Phys. Rev. B* **56**, pp. 8360–8366.

Mokashi, A., Li, S., Wen, B., Kravchenko, S. V., Shashkin, A. A., Dolgopolov, V. T., and Sarachik, M. P. (2012). *Phys. Rev. Lett.* **109**, p. 096405.

Müller, M., Strack, P., and Sachdev, S. (2012). *Phys. Rev. A* **86**, p. 023604.

Novoselov, K. S., Jiang, D., Schedin, F., Booth, T. J., Khotkevich, V. V., Morozov, S. V., and Geim, A. K. (2005). *Proc. Natl. Acad. Sci. USA* **102**, 30, pp. 10451–10453.

Okamoto, T., Hosoya, K., Kawaji, S., and Yagi, A. (1999). *Phys. Rev. Lett.* **82**, p. 3875.

Pastor, A. A., and Dobrosavljević, V. (1999). *Phys. Rev. Lett.* **83**, p. 4642.

Peeters, F. M. (1984). *Phys. Rev. B* **30**, pp. 159–165.

Popović, D. (2012). Glassy dynamics of electrons near the metal-insulator transition, in *Conductor-Insulator Quantum Phase Transitions*, pp. 256–295, eds. V. Dobrosavljević, N. Trivedi, and J. M. Valles, Jr. (Oxford University Press).

Popović, D., Fowler, A. B., and Washburn, S. (1997). *Phys. Rev. Lett.* **79**, p. 1543.

Pradhan, N. R., McCreary, A., Rhodes, D., Lu, Z., Feng, S., Manousakis, E., Smirnov, D., Namburu, R., Dubey, M., Walker, A. R. H., Terrones, H., Terrones, M., Dobrosavljevic, V., and Balicas, L. (2015). *Nano Lett.* **15**, 12, pp. 8377–8384.

Pudalov, V. M., Brunthaler, G., Prinz, A., and Bauer, G. (1998). *JETP Lett.* **68**, p. 442.

Pudalov, V. M., D'Iorio, M., Kravchenko, S. V., and Campbell, J. W. (1993). *Phys. Rev. Lett.* **70**, p. 1866.

Punnoose, A., and Finkelstein, A. M. (2005). *Science* **310**, pp. 289–291.

Radisavljevic, B., and Kis, A. (2013). *Nat. Mater.* **12**, pp. 815–820.

Radonjić, M. M., Tanasković, D., Dobrosavljević, V., Haule, K., and Kotliar, G. (2012). *Phys. Rev. B* **85**, p. 085133.

Raičević, I., Jaroszyński, J., Popović, D., Panagopoulos, C. and Sasagawa, T. (2008). *Phys. Rev. Lett.* **101**, p. 177004.

Raičević, I., Popović, D., Panagopoulos, C., and Sasagawa, T. (2010). *Phys. Rev. B* **81**, p. 235104.

Raičević, I., Popović, D., Panagopoulos, C., and Sasagawa, T. (2011). *Phys. Rev. B* **83**, p. 195133.

Raičević, I., Popović, D., Panagopoulos, C., and Sasagawa, T. (2012). *J. Supercond. Nov. Magn.* **25**, pp. 1239–1242.

Roldán, R., Castellanos-Gomez, A., Cappelluti, E., and Guinea, F. (2015). *J. Phys.: Condens. Matter* **27**, 31, p. 313201.

Rosenbaum, T. F., Field, S. B., and Bhatt, R. N. (1989). *Europhys. Lett.* **10**, p. 269.

Rubi, M., and Perez-Vicente, C. (eds.) (1997). Complex behavior of glassy systems, in *Lecture Notes in Physics*, Vol. 492 (Springer, Berlin).

Sachdev, S. (2011). *Quantum Phase Transitions*, 2nd ed. (Cambridge University Press, UK).

Sakr, M. R., Rahimi, M., and Kravchenko, S. V. (2001). *Phys. Rev. B* **65**, p. 041303.

Sarachik, M. P. (1995). Transport studies in doped semiconductors near the metal-insulator transition, pp. 79–104, in *The Metal-Nonmetal Transition Revisited: A Tribute to Sir Nevill Mott*, eds. P. P. Edwards and C. N. Rao (Francis and Taylor Ltd.).

Sarachik, M. P., Simonian, D., Kravchenko, S. V., Bogdanovich, S., Dobrosavljević, V., and Kotliar, G. (1998). *Phys. Rev. B* **58**, p. 6692.

Schmalian, J., and Wolynes, P. G. (2000). *Phys. Rev. Lett.* **85**, p. 836.

Shashkin, A. A., Kravchenko, S. V., and Klapwijk, T. M. (2001). *Phys. Rev. Lett.* **87**, p. 266402.

Shi, X., Logvenov, G., Bollinger, A. T., Božović, I., Panagopoulos, C. and Popović, D. (2012). *Nat. Mater.* **12**, pp. 47–51.

Shi, X., Popović, D., Panagopoulos, C., Logvenov, G., Bollinger, A. T. and Božović, I. (2012). *Physica B* **407**, pp. 1915–1918.

Skinner, B., and Fogler, M. M. (2010). *Phys. Rev. B* **82**, p. 201306.

Skinner, B., and Shklovskii, B. I. (2010). *Phys. Rev. B* **82**, p. 155111.

Smith, R. P., and Stiles, P. J. (1986). *Solid State Commun.* **58**, pp. 511–514.

Spivak, B., Kravchenko, S. V., Kivelson, S. A., and Gao, X. P. A. (2010). *Rev. Mod. Phys.* **82**, pp. 1743–1766.

Stauffer, D., and Aharony, A. (1994). *Introduction to Percolation Theory: Revised Second Edition* (Taylor & Francis, London).

Struik, L. C. E. (1978). *Physical Aging in Amorphous Polymers and Other Materials* (Elsevier, Amsterdam).

Taur, Y., and Ning, T. H. (1999). *Fundamentals of Modern VLSI Devices* (Cambridge University Press, Cambridge).

Thakur, J. S., and Neilson, D. (1996). *Phys. Rev. B* **54**, p. 7674.

Thakur, J. S., and Neilson, D. (1999). *Phys. Rev. B* **59**, p. R5280.

Tracy, L. A., Hwang, E. H., Eng, K., Ten Eyck, G. A., Nordberg, E. P., Childs, K., Carroll, M. S., Lilly, M. P., and Das Sarma, S. (2009). *Phys. Rev. B* **79**, p. 235307.

Vitkalov, S. A., Zheng, H., Mertes, K. M., and Sarachik, M. P. (2000). *Phys. Rev. Lett.* **85**, p. 2164.

Wang, Q. H., Kalantar-Zadeh, K., Kis, A., Coleman, J. N., and Strano, M. S. (2012). *Nano Nanotechnol.* **7**, pp. 699–712.

Washburn, S., Kim, N. J., Feng, X. G., and Popović, D. (1999a). *Ann. Phys. (Leipzig)* **8**, p. 569.

Washburn, S., Kim, N.-J., Li, K. P., and Popović, D. (1999b). *Mol. Phys. Rept.* **24**, p. 150.

Watanabe, M., Itoh, K. M., Ootuka, Y., and Haller, E. E. (1999). *Phys. Rev. B* **60**, p. 15817.

Widom, A., and Tao, R. (1988). *Phys. Rev. B* **38**, pp. 10787–10790.

Zeng, S. W., Huang, Z., Lv, W. M., Bao, N. N., Gopinadhan, K., Jian, L. K., Herng, T. S., Liu, Z. Q., Zhao, Y. L., Li, C. J., Harsan Ma, H. J., Yang, P., Ding, J., Venkatesan, T., and Ariando (2015). *Phys. Rev. B* **92**, p. 020503.

Chapter 6

Microscopic Theory of a Strongly Correlated Two-Dimensional Electron Gas

M. V. Zverev[a,b] and V. A. Khodel[a,c]

[a]*National Research Center Kurchatov Institute, Moscow 123182 Russia*
[b]*Moscow Institute of Physics and Technology, Dolgoprudny, Moscow District 141700, Russia*
[c]*McDonnell Center for the Space Science and Department of Physics, Washington University, St.Louis, MO 63130, USA*
Zverev_MV@nrcki.ru

6.1 Introduction

A 2D electron gas is a physical model aimed to describe properties of electron systems at a semiconductor–dielectric interface in metal-oxide-semiconductor field-effect transistors (MOSFETs) or at the interface between two different semiconductors in heterostructures, like Si/SiGe, AlGaAs/GaAs, and AlAs/AlGeAs, where electron motion is quantized in a direction perpendicular to the surface. In these systems, the electron density n can be changed by several orders of magnitude by imposition of an external electric field that allows one

Strongly Correlated Electrons in Two Dimensions
Edited by Sergey Kravchenko
Copyright © 2017 Pan Stanford Publishing Pte. Ltd.
ISBN 978-981-4745-37-6 (Hardcover), 978-981-4745-38-3 (eBook)
www.panstanford.com

to trace the transition from a weakly correlated regime to a strongly correlated one, which is characterized by a rearrangement of the standard Fermi liquid (FL) spectrum of single-particle excitations $\epsilon(p) = p_F(p - p_F)/m^*$, to trigger non-Fermi-liquid (NFL) behavior at the so-called Wigner–Seitz radius $r_s = a_B/(\pi n)^{1/2} = \sqrt{2}me^2/p_F$ between electrons (with a_B being the Bohr radius, n the electron density, m the bare electron mass, and p_F the Fermi momentum), whose value is in excess of unity.

The experimental investigations of a 2D interacting electron gas that began almost half a century ago (Fang and Stiles, 1968) gave rise to a fundamental discovery, made almost simultaneously by several groups of experimentalists in the silicon MOSFETs around 15 years ago (Kravchenko and Sarachik, 2004; Pudalov et al., 2002; Shashkin et al., 2002; Shashkin, 2005). In the strongly correlated regime that sets in near a critical density $n_c = 0.8 \times 10^{11} \text{cm}^{-2}$ of the metal–insulator transition, corresponding to $r_s \simeq 8$, the electron effective mass m^* turns out to diverge. Noteworthy, the divergence of the effective mass was observed also in liquid ^{3}He films (Bäuerle et al., 1998; Casey et al., 1998; Neumann et al., 2007) and in a number of heavy-fermion compounds (Gegenwart et al., 2008; Löhneysen et al., 2007). Since available experimental data on divergence of m^* in Si-MOSFETs with quite different impurity concentrations (Kapustin et al., 2009; Shashkin, 2005) give approximately the same result for the corresponding value of the critical density n_c, as in ultralow-disorder quantum wells (Melnikov et al., 2014), we infer that the divergence of m^* is an interaction-induced effect, and impurities are irrelevant to its origin.

Beyond the point where the effective mass diverges, the Landau theory of FLs, being the foundation of condensed matter physics during more than 50 years, fails. That is why the divergence of m^*, associated with the occurrence of NFL behavior of FLs in the strongly correlated regime, is of paramount interest. Alas, the random phase approximation (RPA), served for the description of properties of electron systems for a long time, turns out to be inapplicable to the analysis of the strongly correlated regime, as the comparison of the RPA and Monte Carlo (MC) results for the ground-state energies of a 2D electron gas and linear response functions demonstrates (see below).

In this article, we present the theory of a 2D electron gas based on an ab initio functional approach to the many-body problem developed by Khodel et al. (1994) and Shaginyan (1998). In Section 6.2, the detailed structure of the theory and scheme of the numerical solution of relevant equations (Borisov and Zverev, 2005) are discussed. Within the jellium model, we calculate the ground-state energy and response function versus r_s and compare the results obtained with the MC ones. The method of evaluation of the single-particle spectrum $\epsilon(p)$ of a 2D electron gas is discussed in Section 6.3. We show that within this approach, the divergence of the effective mass occurs at $r_s \simeq 7$. In Section 6.4, we examine the behavior of a 2D electron gas in external magnetic fields, focusing on the disappearance of magnetic oscillations in electron systems of MOSFETs due to merging of spin- and valley-split Landau levels in silicon, being the topological rearrangement of the Landau state, analogous to the formation of the fermion condensate.

6.2 Ab initio Evaluation of the Ground-State Energy and Response Function of a 2D Electron Gas

We address the problem of a 2D electron gas with the pair interaction potential $V(q) = 2\pi e^2/q$, employing the so-called jellium model where ions are treated as an inert positive background that ensures overall charge neutrality. Following Khodel et al. (1994) and Zverev et al. (1996), we begin with writing the Feynman–Hellmann formula (Pines and Noziéres, 1966)

$$e^2 \frac{\partial E_0}{\partial e^2} = \langle \Phi_0 | \hat{H}_{\text{int}} | \Phi_0 \rangle, \tag{6.1}$$

where E_0 is the ground-state energy of the system, Φ_0 denotes the ground-state wave function, and $\hat{H}_{\text{int}}$ stands for the interaction Hamiltonian. We integrate Eq. 6.1 over the coupling constant e^2 from zero to the actual value of the squared electron charge and E_0 is then presented as a sum

$$E_0 = \tau + W_0, \tag{6.2}$$

where τ is energy of gas of noninteracting particles, that is, $\tau = E\,(e^2 = 0)$, while W_0 is the interaction energy presented as a sum

$$W_0 = W_H + W_F + W_c. \tag{6.3}$$

In the conventional jellium model, the Hartree contribution W_H cancels with that of the uniform background of the positive charge. The Fock term W_F, given by

$$W_F = \frac{e^2 n}{2} \int \frac{d^2 q}{2\pi q} \left[S_0(\mathbf{q}) - n\delta(\mathbf{q}) - 1\right] \tag{6.4}$$

is expressed through the static form factor (Pines and Noziéres, 1966) of the 2D noninteracting electron gas

$$S_0(\mathbf{k}) = \frac{2}{n} \int \frac{d^2 p}{(2\pi)^2} n(\mathbf{p}) \left[1 - n(\mathbf{p}+\mathbf{k})\right], \tag{6.5}$$

where $n(p) = \theta(p_F - p)$ is the noninteracting-Fermi-gas momentum distribution.

The remaining contribution, that is, the correlation energy W_c is expressed in terms of the density–density linear response function χ as follows (Pines and Noziéres, 1966):

$$W_c = -\frac{1}{2} \int\limits_0^{e^2} de^2 \int \frac{d^2 q}{2\pi q} \int\limits_0^\infty \frac{d\omega}{\pi} \, \text{Im} \left[\chi\,(\mathbf{q}, \omega) - \chi_0(\mathbf{q}, \omega)\right]. \tag{6.6}$$

The explicit formula for χ reads

$$\chi\,(\mathbf{k}, \omega) = \frac{\delta n(\mathbf{k}, \omega)}{\delta V_0(\mathbf{k}, \omega)} \tag{6.7}$$

where $\delta V_0(\mathbf{k}, \omega)$ is the Fourier transform of the variation of a spatial- and time-dependent external field $V_0(\mathbf{r}, t)$, and $\delta n(\mathbf{k}, \omega)$ is the same of the variation $\delta n(\mathbf{r}, t)$ of the density in this field. Correspondingly,

$$\chi_0(\mathbf{k}, \omega) = 2 \int \frac{d^2 p}{(2\pi)^2} \frac{n(\mathbf{p}) - n(\mathbf{p}-\mathbf{k})}{\epsilon_{\mathbf{p}}^0 - \epsilon_{\mathbf{p}-\mathbf{k}}^0 - \omega}, \tag{6.8}$$

with $\epsilon_{\mathbf{p}}^0 = p^2/2m - \mu$, is the linear response function of the 2D noninteracting electron gas.

The response functions χ and χ_0 are connected with each other in terms of the effective interaction R, introduced as follows: (Khodel et al., 1994)

$$\chi^{-1}(\mathbf{k}, \omega) = \chi_0^{-1}(\mathbf{k}, \omega) - R(\mathbf{k}, \omega), \tag{6.9}$$

or, equivalently, by the relation

$$\chi(\mathbf{k}, \omega) = \frac{\chi_0(\mathbf{k}, \omega)}{1 - R(\mathbf{k}, \omega)\chi_0(\mathbf{k}, \omega)}. \tag{6.10}$$

To obtain an equation for the quantity R we employ the connection between the second variational derivative of E_0 and the linear response function of the system (Khodel et al., 1994),

$$\frac{\delta^2 E_0}{\delta n(\mathbf{k}, \omega)\delta n(-\mathbf{k}, -\omega)} = -\chi^{-1}(\mathbf{k}, \omega). \tag{6.11}$$

Since

$$\frac{\delta^2 \tau}{\delta n(\mathbf{k}, \omega)\delta n(-\mathbf{k}, -\omega)} = -\chi_0^{-1}(\mathbf{k}, \omega), \tag{6.12}$$

we immediately arrive at the variational connection between the effective interaction and the interaction energy that reads

$$R(\mathbf{k}, \omega) = \frac{\delta^2 W_0}{\delta n(\mathbf{k}, \omega)\delta n(-\mathbf{k}, -\omega)}. \tag{6.13}$$

The effective interaction R can be separated into the exchange term

$$R_{\text{ex}}(\mathbf{k}, \omega) = \frac{\delta^2 W_F}{\delta n(\mathbf{k}, \omega)\delta n(-\mathbf{k}, -\omega)} \tag{6.14}$$

that can be calculated in the closed form, and the correlation contribution

$$R_c(\mathbf{k}, \omega) = \frac{\delta^2 W_c}{\delta n(\mathbf{k}, \omega)\delta n(-\mathbf{k}, -\omega)}. \tag{6.15}$$

Combining this relation with Eq. 6.6 we arrive at a closed functional equation for the quantity of our interest, the effective interaction (Khodel et al., 1994):

$$R(\mathbf{k}, \omega) = \frac{2\pi e^2}{k} + R_{\text{ex}}(\mathbf{k}, \omega) - \frac{1}{2}\frac{\delta^2}{\delta n(\mathbf{k}, \omega)\delta n(-\mathbf{k}, -\omega)}$$

$$\times \int_0^{e^2} de^2 \int \frac{d^2 q}{2\pi q} \int_0^{\infty} \frac{d\omega'}{\pi} \operatorname{Im}\left[\chi(\mathbf{q}, \omega') - \chi_0(\mathbf{q}, \omega')\right]. \tag{6.16}$$

To calculate the variational derivatives on the right-hand side of Eq. 6.16, we follow the method proposed by Khodel et al. (1994)

and Shaginyan (1998) where these derivatives are evaluated within a local approximation (Kawazoe et al., 1977):

$$\frac{\delta^2 \chi(\mathbf{q}, \omega)}{\delta n(\mathbf{k}, \varepsilon)\delta n(-\mathbf{k}, -\varepsilon)} \simeq \frac{1}{2} \left[\frac{d^2 \chi(\mathbf{q}+\mathbf{k}, \omega+\varepsilon)}{dn^2} + \frac{d^2 \chi(\mathbf{q}-\mathbf{k}, \omega-\varepsilon)}{dn^2} \right].$$
(6.17)

The local approximation proved to be effective in evaluation of the ground-state energies of 3D electron gas and neutron matter (Khodel et al., 1994). Within this approximation, the effective interaction R becomes frequency independent

$$R(k) = \frac{2\pi e^2}{k} + R_{ex}(k) + R_c(k).$$
(6.18)

The exchange term $R_{ex}(k)$ is calculated through the static form factor (Eq. 6.5), as follows (Borisov and Zverev, 2005):

$$R_{ex}(k) = -\frac{2e^2}{\pi p_F^2} \int_0^{2p_F} \frac{q^2 dq}{\sqrt{4p_F^2 - q^2}} \frac{K(4qk/(q+k)^2)}{q+k},$$
(6.19)

where $K(z)$ is an elliptical integral of the first kind.

To calculate the second term in Eq. 6.18 we rotate the integration contour to perform integration along the imaginary axis:

$$R_c(k) = -\frac{1}{2}\frac{d^2}{dn^2} \int_0^{e^2} de^2 \int \frac{d^2 q}{2\pi |\mathbf{q}-\mathbf{k}|} \int_{C_I} \frac{d\omega}{2\pi i} \chi_0(q, \omega)$$

$$\times \left(\frac{1}{1 - R(q)\chi_0(q, \omega)} - 1 \right),$$
(6.20)

where an explicit expression for the response function χ_0 at $\omega = iw$ is given by

$$\chi_0(t, z) = -\frac{m}{\pi} \left[1 - \frac{1}{\sqrt{2}t} \left(\sqrt{(1+z^2-t^2)^2 + 4t^2 z^2} - 1 - z^2 + t^2 \right)^{1/2} \right]$$
(6.21)

with $t = k/2p_F$, $z = mw/kp_F$.

The next iteration step is to insert the effective interaction

$$R_0(q) = V(q) + R_{ex}(q)$$
(6.22)

into the right-hand side of Eq. 6.20. Then upon integrating with respect to e^2, we obtain

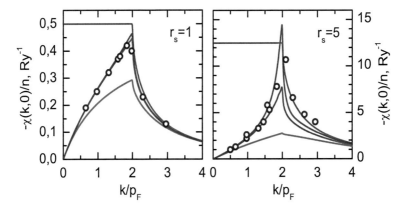

Figure 6.1 Response function of the 2D electron gas in the static limit, $-\chi(k, 0)/n$, for $r_s =$ (left) 1 and (right) 5. The MC results (Moroni et al., 1992) are represented by brown circles, the violet line shows the response function $-\chi_0(k, 0)/n$ (Stern et al., 1967), and the green line represents the RPA calculation. The blue line shows the calculation including only the correction $R_{ex}(k)$, and the red line additionally takes into account the correction $R_c(k)$. From Borisov, V. V., and Zverev, M. V. (2005). *JETP Letters* **81**, 10, pp. 503–508. With permission of Springer.

$$R_c(k) = \frac{e^2}{2} \frac{d^2}{dn^2} \int \frac{d^2q}{2\pi} \frac{1}{R_0(q)|\mathbf{q}-\mathbf{k}|} \int_{C_I} \frac{d\omega}{2\pi i}$$

$$\times \left(R_0(q)\chi_0(q, \omega) + \ln[1-R_0(q)\chi_0(q, \omega)] \right), \quad (6.23)$$

The accuracy of this procedure, employing only one iteration in evaluation of the correlation correction, can be estimated by comparing theoretical characteristics of the 2D electron gas with results of MC calculations. Figure 6.1 shows the response function in the static limit, $-\chi(k, 0)$, divided by the electron density n for $r_s = 1$ and 5. One can see that the inclusion of the exchange correction considerably improves the initial RPA calculations. For illustration, in Fig. 6.2 the calculated correlation energy of the 2D electron gas is compared with the corresponding MC data (Kwon et al., 1993). As seen, the RPA does overestimate the absolute value of the energy by approximately the factor of 2. A major part of this discrepancy is eliminated by taking into account the exchange correction, and the inclusion of the correlation correction further reduces the discrepancy to 12% or less.

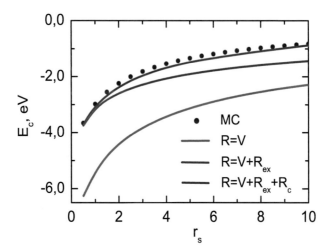

Figure 6.2 Correlation energy of the 2D electron gas. The brown circles represent the data of MC simulations (Kwon et al., 1993), the green line is the RPA calculation, the blue line shows our calculation taking into account the term $R_{ex}(k)$ of the effective interaction, and the red line is the calculation with $R(k) = V(k) + R_{ex}(k) + R_c(k)$. From Borisov, V. V., and Zverev, M. V. (2005). *JETP Letters* **81**, 10, pp. 503–508. With permission of Springer.

6.3 Ab initio Evaluation of Single-Particle Excitations of a 2D Electron Gas

The coincidence of the Landau quasiparticle momentum distribution with that of noninteracting particles $n(p) = \theta(p_F - p)$ allows one to consider Eqs. 6.2–6.6 as an *explicit form* of the Landau functional $E_0[n(p)]$. Further, in Landau theory, the single-particle spectrum $\epsilon(p)$ is just a variational derivative of E_0 with respect to the distribution $n(p)$ (Landau, 1956, 1959; Lifshits and Pitaevski, 1980)

$$\epsilon(\mathbf{p}) = \frac{\delta E_0}{\delta n(\mathbf{p})}. \tag{6.24}$$

Therefore the knowledge of the explicit form of $E_0[n(p)]$, given by Eqs. 6.2–6.6, allows one to perform a parameter-free evaluation of the electron spectrum $\epsilon(p)$ of a 2D electron gas without adjustable parameters. Straightforward calculations yield

$$\frac{\delta\chi(\mathbf{q},\omega)}{\delta n(\mathbf{p})} = \varphi^2(\mathbf{q},\omega)\frac{\delta\chi_0(\mathbf{q},\omega)}{\delta n(\mathbf{p})} + \chi^2(\mathbf{q},\omega)\frac{\delta R(\mathbf{q})}{\delta n(\mathbf{p})}, \qquad (6.25)$$

where

$$\varphi(\mathbf{q},\omega) = [1 - R(\mathbf{q})\chi_0(\mathbf{q},\omega)]^{-1} \qquad (6.26)$$

that stems from Eq. 6.10. Then one finds

$$\epsilon(p) = \epsilon_p^0 + \epsilon_1(p) + \epsilon_2(p), \qquad (6.27)$$

with

$$\epsilon_1(p) = -\frac{1}{2}\int\frac{d^2q}{2\pi}\frac{e^2}{|\mathbf{p}-\mathbf{q}|}$$

$$-\frac{1}{2}\int_0^{e^2}de^2\int\frac{d^2q}{2\pi q}\int_0^{\infty}\frac{d\omega}{\pi}\mathrm{Im}\left[\varphi^2(\mathbf{q},\omega)\frac{\delta\chi_0(\mathbf{q},\omega)}{\delta n(\mathbf{p})}\right], \quad (6.28)$$

and

$$\epsilon_2(p) = -\frac{1}{2}\int_0^{e^2}de^2\int\frac{d^2q}{2\pi q}\int_0^{\infty}\frac{d\omega}{\pi}\mathrm{Im}\left[\chi^2(\mathbf{q},\omega)\frac{\delta R(q)}{\delta n(\mathbf{p})}\right]. \qquad (6.29)$$

The variation of $\chi_0(\mathbf{q},\omega)$ with respect to $n(p)$ is given by

$$\frac{\delta\chi_0(\mathbf{q},\omega)}{\delta n(p)} = \frac{n(\mathbf{p}-\mathbf{q})}{\epsilon_p^0 - \omega - \epsilon_{p-q}^0 - i\delta} + \frac{1 - n(\mathbf{p}-\mathbf{q})}{\epsilon_p^0 - \omega - \epsilon_{p-q}^0 + i\delta}$$

$$+ \frac{n(\mathbf{p}-\mathbf{q})}{\epsilon_p^0 + \omega - \epsilon_{p-q}^0 - i\delta} + \frac{1 - n(\mathbf{p}-\mathbf{q})}{\epsilon_p^0 + \omega - \epsilon_{p-q}^0 + i\delta}. \qquad (6.30)$$

With this result, we are led to

$$\epsilon_1(p) = -\frac{2e^2}{\pi}p + \int_0^{e^2}de^2\int_{p_F}^{p}\frac{d^2q}{2\pi|\mathbf{p}-\mathbf{q}|}\varphi^2(\mathbf{q},\epsilon_p^0-\epsilon_q^0)$$

$$+ \int_0^{e^2}de^2\int\frac{d^2q}{2\pi q}\int_{-\infty}^{\infty}\frac{dw}{2\pi}\left(\varphi^2(q,iw)-1\right)\frac{\epsilon_p^0-\epsilon_{p-q}^0}{(\epsilon_p^0-\epsilon_{p-q}^0)^2+w^2}. \qquad (6.31)$$

Once the effective interaction $R(k)$ is found, evaluation of the term $\epsilon_1(p)$ in Eq. 6.28 reduces to calculating integrals. As for Eq. 6.29 for finding $\epsilon_2(p)$, it contains not only $R(k)$ but also the variational derivative $\delta R(k)/\delta n(p)$, a linear integral equation for which was derived by Zverev et al. (1996). In principle, the corresponding equation can be obtained in the 2D case as well. However, for values of $r_s \lesssim 10$, considered below, the contribution of $\epsilon_2(p)$ to the spectrum $\epsilon(p)$ turns out to be rather small, and therefore, following the strategy of Zverev et al. (1996), we will apply a version of the local approximation adapted to evaluation of variational derivatives of the linear response function. In this case, upon inserting Eq. 6.5 into Eq. 6.4 and after some manipulations we obtain

$$\frac{\delta R_{ex}(k)}{\delta n(p)} = -\frac{1}{2} \int \frac{d^2q}{(2\pi)^2} \frac{2\pi e^2}{|\mathbf{q}-\mathbf{k}|} \int_0^\infty \text{Im} \frac{\delta}{\delta n(p)} \frac{d^2 \chi_0(\mathbf{q},\omega)}{dn^2} \frac{d\omega}{\pi}.$$

(6.32)

Bearing in mind that $d/dn = (\pi/p_F)d/dp_F$, we find

$$\frac{\delta}{\delta n(p)} \frac{d^2 \chi_0(\mathbf{k},\omega)}{dn^2} = \frac{i\pi^2}{p_F} \int dO_\mathbf{p} dO_{\mathbf{p_1}} dp_1 \delta'(p_1 - p_F)\delta(\mathbf{p_1}-\mathbf{k}-\mathbf{p})$$

$$\times \Big[\delta(\epsilon_{\mathbf{p_1}}-\epsilon_{\mathbf{p_1}+\mathbf{k}}-\omega) + \delta(\epsilon_{\mathbf{p_1}}-\epsilon_{\mathbf{p_1}+\mathbf{k}}+\omega)\Big].$$

(6.33)

Upon inserting this expression into Eq. 6.32 and evaluating integrals we arrive at

$$\frac{\delta R_{ex}(k)}{\delta n(p)} = -\frac{\pi e^2}{2p_F^3} \int_0^{2\pi} \left\{ -\frac{4}{(l+1)^2} K\left(\frac{2\sqrt{l}}{l+1}\right) \right.$$

$$\left. +\frac{2}{l+1} \left[\frac{l+1}{l-1} E\left(\frac{2\sqrt{l}}{l+1}\right) + \frac{l-1}{l+1} K\left(\frac{2\sqrt{l}}{l+1}\right)\right] \right\} d\vartheta, \quad (6.34)$$

where $l = \sqrt{p^2+k^2-2pk\cos\vartheta}/p_F$ and $E(z)$ is an elliptical integral of the second kind. We substitute the variational derivative found above into Eq. 6.29 for $\epsilon_2(p)$, in which we approximate the effective interaction by $R_0(q)$. The integral with respect to the coupling constant is calculated analytically. As a result, we obtain

$$\epsilon_2(p) = -\frac{e^2}{2} \int \frac{d^2q}{2\pi} \frac{1}{|\mathbf{p}-\mathbf{q}|} \frac{1}{R_0^2(q)} \frac{\delta R_{ex}(q)}{\delta n(p)}$$

$$\times \int_{-\infty}^\infty \frac{dw}{2\pi} \left[\frac{R_0(q)\chi_0(q,iw)}{1-R_0(q)\chi_0(q,iw)} + \ln\Big(1-R_0(q)\chi_0(q,iw)\Big)\right].$$

(6.35)

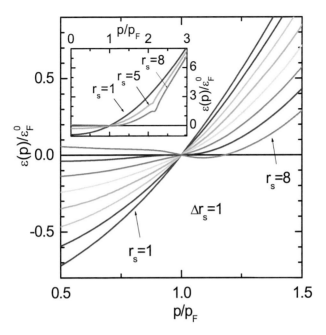

Figure 6.3 Single-particle spectra of the 2D electron gas, $\epsilon(p)$, in units of $\epsilon_F^0 = p_F^2/2m$ for the values of the parameter r_s from 1 to 8 with the step $\Delta r_s = 1$. The inset shows the spectra for $r_s = 1$, 5, and 8 in the momentum interval $0 < p/p_F < 3$. From Borisov, V. V., and Zverev, M. V. (2005). *JETP Letters* **81**, 10, pp. 503–508. With permission of Springer.

The single-particle spectra $\epsilon(p)$ calculated from Eq. 6.27, where $\epsilon_1(p)$ and $\epsilon_2(p)$ are determined by Eqs. 6.31 and 6.35 are shown in Fig. 6.3 for the values of r_s from 1 to 8 with the step $\Delta r_s = 1$. As r_s increases, the spectrum flattens out in the region of momenta $p < p_F$, and, at $r_s \simeq 7$, the group velocity $v_F = d\epsilon(p)/dp)_{p=p_F}$ changes sign. A critical point where v_F vanishes is the point of the topological instability of the Landau state, since at this point no one symmetry inherent in the Landau state breaks down. The inset in Fig. 6.3 shows the spectra for $r_s = 1$, 5, and 8 in a wider interval of momenta: $0 < p/p_F < 3$. Noteworthy, irregularities in electron spectra near $p \approx 2p_F$ that become more pronounced as r_s increases are associated with the plasmon contribution to the second term of Eq. 6.31. However, as is seen in Fig. 6.3, in the region $r_s \lesssim 7$, where the flattening of the spectrum near the Fermi surface

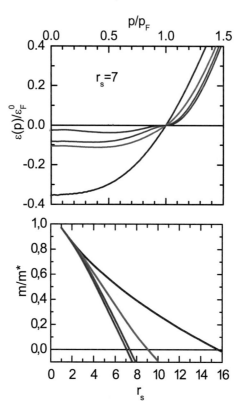

Figure 6.4 Top: Single-particle spectra $\epsilon(p)$ of the 2D electron gas measured from the chemical potential in different approximations for r_s = 7. The brown line shows the RPA calculation, the green line represents the term $\epsilon_1(p)$ calculated with the effective interaction $R_0(k)$, the blue line shows the spectrum $\epsilon_1(p)$ calculated with $R(k) = R_0(k) + R_c(k)$, and the red line represents the spectrum with the term $\epsilon_2(p)$. Bottom: The ratio m/m^* in different approximations versus the parameter r_s. The color legend for the curves is the same as in the top panel. From Borisov, V. V., and Zverev, M. V. (2005). *JETP Lett.* **81**, 10, pp. 503–508. With permission of Springer.

clearly manifests itself, these irregularities are still not involved in the formation of the spectrum instability away from p_F.

Single-particle spectra calculated within different approxima- tions at r_s = 7 are displayed in the top panel of Fig. 6.4. The approximation of the effective interaction $R(k)$ by the bare one $V(k) = 2\pi e^2/k$ in Eq. 6.28 for the term $\epsilon_1(p)$ corresponds to the

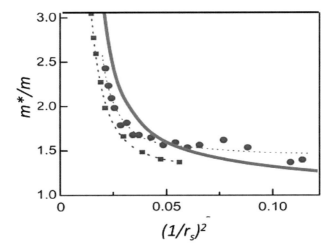

Figure 6.5 Ratio m^*/m calculated for the 2D electron gas versus the parameter $1/r_s^2$ (green line). Red and blue circles show the data for the ratio of the effective mass to the band mass in (100) silicon (Shashkin et al., 2002) and (111) silicon (Kapustin et al., 2009) MOSFETs, respectively.

RPA. A detailed analysis of this case was done by Zhang et al. (2005). The RPA effective mass evaluated at $r_s = 7$ is about 2.0 (see the bottom panel of Fig. 6.4 where the inverse of the effective mass $m/m^* = (m/p_F)(d\epsilon(p)/dp)_{p=p_F}$ is displayed). In this approximation the electron effective mass m^* diverges at $r_s \simeq 16$. Accounting for the exchange term in the effective interaction $R(k) = 2\pi e^2/k + R_{ex}(k)$ yields a considerable contribution to the spectrum $\epsilon_1(p)$ already at $r_s = 7$ (green line in the top panel of Fig. 6.4): the value of $m^*(r_s = 7)$ increases by factor 2 compared to its RPA value, while the function $m^*(r_s)$ diverges, as seen from the bottom panel, at $r_s \simeq 9$ instead of the RPA value 16.

The magnitude of the contribution to the spectrum $\epsilon(p)$, related to the correlation correction in the effective interaction $R(k) = 2\pi e^2 + R_{ex}(k) + R_c(k)$ (blue line), is seen to be much smaller than the contribution due to the exchange correction (green line). However, due to marked irregularities in the vicinity of $p = p_F$, the former contribution considerably changes the effective mass which diverges at $r_s \simeq 7.3$. The magnitude of the term $\epsilon_2(p)$ is not small; however, near the Fermi surface, it changes slowly, so its contribution reduces mainly to renormalization of the chemical potential. As a result, the group velocity $v(p) = \partial\epsilon(p)/\partial p$ changes sign at the Fermi surface, shifting from $r_s \simeq 7.3$ to $r_s \simeq 7$ the value of r_s, at which the Fermi velocity $v_F(r_s)$ changes sign and the topological stability of the Landau state is violated (see the bottom panel of Fig. 6.4). The inclusion of this contribution does not change the character of the instability: the latter arises at the point p_F, that is, as the divergence of the effective mass. This can be seen from the top panel of Fig. 6.4, where the red line represents the spectrum calculated with allowance for all contributions at $r_s = 7$. The calculation shows that the first and second derivatives of the spectrum simultaneously become equal to zero at the point p_F; therefore, near the Fermi surface, we have $\epsilon(p) \propto (p-p_F)^3$. We also note that because of the flatness of the spectrum (within the interval $0 < p < p_F$, its absolute value does not exceed $0.05\,\epsilon_F^0$), the corrections ignored by us (the corrections associated, for example, with the subsequent iteration steps in solving Eqs. 6.18–6.20 or with the change from the local approximation to the exact calculation of variational derivatives) could shift the point of formation of instability to the region of momenta $p < p_F$.

It is interesting to compare the results discussed above with those obtained along the same lines for 3D electron gas by Zverev et al. (1996). On the one hand, they are rather different quantitatively. In the 3D case the effective mass diverges at a very large $r_s \simeq 23$; moreover, instability owing to the arising of a new zero of the spectrum $\epsilon(p)$ at $p_b \simeq 0.6\,p_F$ occurs at $r_s \simeq 21$, that is, in advance to the point of the divergence of the effective mass. On the other hand, in both cases the spectra $\epsilon(p)$ possess a common qualitative feature: they are very flat in the whole region $0 < p < p_F$ (compare Fig. 6.3 with Fig. 3 of Zverev et al., 1996).

To conclude this section we compare the calculated effective mass with data on (100) and (111) Si-MOSFETs (Kapustin et al., 2009; Shashkin et al., 2002) presented in Fig. 6.5 as a ratio of the effective mass to the band mass versus $1/r_s^2$. Our theoretical result shown by a green curve qualitatively reasonably agrees with the experimental data.

6.4 Disappearance of de Haas–van Alphen and Shubnikov–de Haas Magnetic Oscillations in MOSFETs as the Precursor of a Topological Rearrangement of the Landau State

The density n_c where the electron effective mass m^* diverges is a critical point beyond which the standard FL theory, with the Landau spectrum $\epsilon(p) = p_F(p - p_F)/m^*$, fails. Importantly, no one symmetry inherent in the Landau state breaks down, so the rearrangement of the Landau state at this point is a *topological transition*. To gain insight into the topological structure of a new ground state, emerging beyond this point, it is instructive to analyze the so-called Landau-level fan diagram, containing experimental data on Shubnikov–de Haas (SdH) oscillations of the resistivity of electron systems of MOSFETs in perpendicular magnetic fields B *before the critical point is reached*. This diagram is presented in Fig. 6.9, taken from Kravchenko et al. (2000). One of its salient features, which cannot be explained by accustomed broadening of Landau levels due to impurity scattering, is that near the critical density n_c, oscillation minima, corresponding to the filling factors $v = 4k + 4$ disappear that results in doubling of oscillation periods, because minima $v = 4k+2$ persist down to the critical density n_c where all the oscillations vanish. This behavior is puzzling, since at high electron densities, it is the cyclotron minima with $v = 4, 8, 12, 16, \ldots$ that dominate in the oscillation pattern (Kravchenko et al., 2000). Importantly, this behavior cannot be accounted by the impurity broadening of quantum levels (Shashkin et al., 2014).

As we will see, the resolution of this puzzle may provide valuable information on possible routes of the rearrangement of the ground state beyond the critical point where the electron effective mass of 2D homogeneous electron liquid diverges, and the Landau state is necessarily rearranged. To clarify this statement several preliminary remarks are in order. First of all, at $T = 0$ by virtue of the twofold valley degeneracy, imposition of the perpendicular magnetic field B on the 2D electron system of MOSFETs creates four subsystems of Landau levels. If, for the sake of simplicity, the spin and valley degeneracy is neglected, we are then left with four degenerate sets of Landau levels, with the cyclotron frequency $\omega_c = eB/m^*c$ that determines energy splitting of neighbor Landau levels.

Given the electron density n, variations of B around $B(N) = 2\pi cn/eN$ where N is the number of the last fully occupied quantum level, whose value is large in weak magnetic fields, lead to a redistribution of electrons in the levels closest to the Fermi level. At $B > B(N)$, the density of quasiparticles in the last filled level, whose energy is denoted further by ϵ_k, decreases by $Np(B - B(N))$, where $p = e/2\pi c$ is the degeneracy factor. In the opposite case of $B < B(N)$, the density of quasiparticles in the nearest to the Fermi surface empty level, whose energy is ϵ_{k+1}, increases by $Np(B(N) - B)$. When calculating the corresponding ground-state energy variation $E(B) - E(B(N))$ we will ignore the variation of energy of completely filled quantum levels, since it changes continuously with B. An irregular part of the difference $E(B) - E(B(N))$ is associated with contributions coming from extra quasiparticles (or quasiholes). It is evaluated with the aid of FL formula $\delta E = \sum \epsilon_\lambda \delta n_\lambda$, where λ includes all quantum numbers specifying single-particle levels. In the case at issue, one finds

$$\delta E(B < B(N)) = Np(B(N) - B)\epsilon_{k+1},$$
$$\delta E(B > B(N)) = Np(B(N) - B)\epsilon_k. \tag{6.36}$$

We see that at $T = 0$ the magnetic moment $M = -dE/dB$ experiences jumps

$$\Delta M(B(N)) = dE(B \rightarrow B(N) - 0)/dB - dE(B \rightarrow B(N) + 0)/dB, \tag{6.37}$$

at fields $B = B(N)$, for which filling of the Landau level is complete, the jump magnitude being given by

$$\Delta M(B(N)) = Np(\epsilon_{k+1}(B(N)) - \epsilon_k(B(N))). \tag{6.38}$$

Thus we infer that in correlated 2D electron systems, the $T = 0$ jumps of the magnetic moment M occur at the same points, as in a noninteracting electron gas; however, their magnitude is renormalized by e–e interactions due to the renormalization of the electron effective mass. Insertion of the cyclotron splitting into Eq. 6.38 does provide the result

$$\Delta M(B(N)) = en/m^*c. \tag{6.39}$$

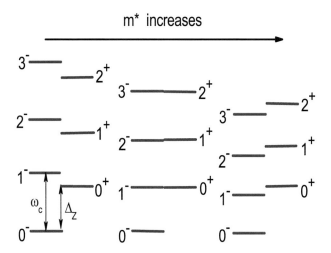

Figure 6.6 Spin-up (red) and spin-down (blue) Landau levels for a 2D electron system of a MOSFET in a magnetic field. The cyclotron frequency $\omega_c = eB/m^*c$ decreases with increasing m^*, while the spin splitting Δ_Z remains almost unchanged. As a result level crossing must occur.

In a noninteracting electron gas where $m^* = m$, this formula coincides with the textbook formula (Abrikosov, 1988).

Spin splitting $\Delta_Z = g\mu_B B$, where $\mu_B = e/2mc$ is the Bohr magneton, and $g \simeq 1.4\,g_0$, the medium Lande factor, with $g_0 = 2$ (Abrahams et al., 2001; Dolgopolov, 2007; Kravchenko and Sarachik, 2004; Mokashi et al., 2012; Shashkin, 2005), comes into play to change the structure of quantum levels and thereby produce some disturbance of the above textbook pattern of magnetic oscillations at low electron densities where the cyclotron splitting becomes suppressed, since the effective mass $m^*(n)$ diverges toward the metal–insulator transition density $n_c \simeq 8 \times 10^{10}$ cm^{-2} as $m/m^* \simeq (n - n_c)/n$, while the spin splitting Δ_Z remains almost unchanged. This implies that at some density near n_c, the Landau level 1^- must cross the level 0^+, the level 2^- should intersect the level 1^+, etc. (see Fig. 6.6).

The level crossing occurs at a density n_{cr} where the difference

$$D(B) = \epsilon_{k+1}^-(B) - \epsilon_k^+(B) = \omega_c(B) - \Delta_Z(B), \qquad (6.40)$$

being positive at high electron densities, due to the smallness of the ratio $m_b/m = 0.19$, (m_b stands for the band electron mass

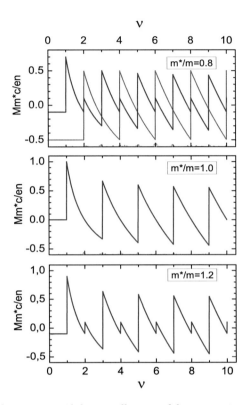

Figure 6.7 de Haas–van Alphen oscillations of the magnetic moment in the simplified model (Abrikosov, 1988). Magnetic moment in units of en/m^*c is given versus $\nu = n/n_0 B$ (red lines) at $m^*/m = 0.2$ (top), $m^*/m = 1.0$ (middle), and $m^*/m = 1.2$ (bottom). The green line shows dHvA oscillations without spin splitting.

in MOSFETs), changes sign. At this point the corresponding de Haas–van Alphen (dHvA) oscillation vanishes. However, at any other point of the phase diagram, the original beat pattern of the dHvA oscillations is recovered, so in the standard FL formalism, the number of jumps exhibited by M coincides with the number of the completely filled Landau levels that remain unaffected by the level crossing (for an illustration see Fig. 6.7), and therefore within the standard FL pattern, the disappearance of any oscillation minimum is *impossible*.

The FL scenario discussed so far is based on the conventional wisdom that quantum levels may merely cross or repel each other. However, in a many-body problem, there exists an additional option: the single-particle levels can *merge with each other* (Khodel et al., 2007), which opens a new avenue of attack on the problem discussed. Indeed, as seen from Eqs. 6.38 and 6.40, the disappearance of magnetic oscillations, observed in the whole region of the phase diagram, implies nullification of the difference D between energies of neighbor quantum levels in this region that makes the merging scenario relevant to the problem. Furthermore, as the analysis performed by Shashkin et al. (2014) has shown, it is merging of those quantum levels, which intersect each other in the standard FL scenario, that explains the long-standing puzzle associated with doubling of periods of dHvA and SdH oscillations in MOSFETs. This result is especially important because, as we will see, the merging phenomenon is reminiscent of swelling of the Fermi surface occurring in homogeneous Fermi systems beyond the critical point where the Landau state loses the topological stability (Khodel and Shaginyan, 1990).

To facilitate the analysis of the merging phenomenon let us address a two-level problem, assuming a pair k^+ and $(k+1)^-$ of quantum levels adjacent to the Fermi surface to be *empty*, with the positive difference $D(B(N)) = \epsilon_{k+1}^-(B(N)) - \epsilon_k^+(B(N))$, whose value is less than the distance to other quantum levels. When B is reduced and therefore the number of quasiparticles associated with completely filled Landau levels declines, the extra quasiparticles populate the empty level k^+. The standard FL pattern of magnetic oscillations suggests that there are two jumps of the magnetic moment: The first occurs when the lower quantum level k^+ becomes completely filled; the second when filling of both the levels is ended. The elucidation of the impact of *e-e* interactions on this pattern is simplified in weak magnetic fields, where the ratio $\delta n/n \simeq 1/N$ is rather small, and variations of the energies ϵ_{k+1}^- and ϵ_k^+ can be evaluated in terms of matrix elements of the electron–electron scattering amplitude Γ using Eq. 6.36 and another Landau relation (Migdal, 1967),

$$\delta\epsilon_\lambda = \sum_l \Gamma_{\lambda,\sigma} \delta n_\sigma. \tag{6.41}$$

Equation 6.41 then reduces to

$$\epsilon_{k+1}^-(B) - \epsilon_{k+1}^-(B(N)) = Np(B(N) - B)\Gamma_{k+1,k}^{-+},$$
$$\epsilon_k^+(B) - \epsilon_k^+(B(N)) = Np(B(N) - B)\Gamma_{k,k}^{+,+}. \qquad (6.42)$$

In deriving Eq. 6.42, the approximation of the equidistance of the Landau spectra, appropriate at weak magnetic fields, is applied. Matrix elements $\Gamma_{k',k}$, entering this equation, result from integration of the initial matrix elements of the static limit of the scattering amplitude over x momentum projections.

Upon subtracting the second of Eq. 6.42 from the first, we are led to

$$D(B) \equiv \epsilon_{k+1}^-(B) - \epsilon_k^+(B) = D(B(N)) - \frac{B(N) - B}{B(N)}\gamma_k^+, \qquad (6.43)$$

where $\gamma_k^+ = n(\Gamma_{k,k}^{++} - \Gamma_{k+1,k}^{-+})$. Here we have employed relation $N = n/pB(N)$. At $\gamma_k^+ < 0$, the right-hand side of Eq. 6.43 stays positive, and the standard pattern of dHvA oscillations persists. In the opposite case $\gamma_k^+ > 0$, however, the level ordering is rearranged in the domain where

$$D(B) \equiv D(B(N)) - \frac{B(N) - B}{B(N)}\gamma_k^+ < 0, \qquad (6.44)$$

triggering resettlement of all the extra quasiparticles to the level $(k+1)^-$. The set of equations in Eq. 6.42 is then replaced by

$$\epsilon_{k+1}^-(B) - \epsilon_{k+1}^-(B(N)) = Np(B(N) - B)\Gamma_{k+1,k+1}^{--},$$
$$\epsilon_k^+(B) - \epsilon_k^+(B(N)) = Np(B(N) - B)\Gamma_{k,k+1}^{+-}, \qquad (6.45)$$

to yield

$$\epsilon_{k+1}^-(B) - \epsilon_k^+(B) = D(B(N)) + \frac{B(N) - B}{B(N)}\gamma_{k+1}^-, \qquad (6.46)$$

with $\gamma_{k+1}^- = n(\Gamma_{k+1,k+1}^{--} - \Gamma_{k,k+1}^{+-})$. In case the right-hand side of this relation has a positive sign, (this case takes place at $\gamma_{k+1}^- > 0$), all the quasiparticles must move back to the level k^+, implying that Eqs. 6.43 and 6.46 are *mutually incompatible*, provided

$$\gamma_k^+ > 0, \qquad \gamma_{k+1}^- > 0. \qquad (6.47)$$

Indeed, according to Eq. 6.44, the level k^+ must be *empty*. If so, the difference $\epsilon_{k+1}^- - \epsilon_k^+$ is to be found from Eq. 6.46. Its value turns out to be *positive*, implying that the original level ordering should be restored, leading to a new round of quasiparticle resettling and implying a vicious logical circle. Thus, in confronting the case (Eq. 6.47), the problem becomes *insoluble*.

The resolution of this inconsistency, first addressed in (Khodel et al., 2007), lies in the mechanism of a topological phase transition associated with *swelling* of the Fermi surface, described in homogeneous matter by the variational condition (Khodel and Shaginyan, 1990)

$$\frac{\delta E}{\delta n(\mathbf{p})} \equiv \epsilon(\mathbf{p}) = \mu, \tag{6.48}$$

where μ is the chemical potential. In systems with a discrete single-particle spectrum, swelling reduces to merging of neighboring single-particle levels adjacent to the Fermi level (Khodel et al., 2007). In the model addressed in this section the equation of merging that stems from Eq. 6.48 takes the simple form

$$\epsilon_{k+1}^-(B) = \epsilon_k^+(B) = \mu. \tag{6.49}$$

Correspondingly, both the levels $(k+1)^-$ and k^+ exhibit *partial occupation*, with filling factors δ_k^+ and δ_{k+1}^- that obey the normalization condition

$$\delta_k^+ + \delta_{k+1}^- = 1 \tag{6.50}$$

and relations

$$\epsilon_{k+1}^-(B) - \epsilon_{k+1}^-(B(v)) = n\frac{B(v) - B}{B(v)}\left(\delta_k^+\Gamma_{k+1,k}^{-+} + \delta_{k+1}^-\Gamma_{k+1,k+1}^{--}\right),$$

$$\epsilon_k^+(B) - \epsilon_k^+(B(v)) = n\frac{B(v) - B}{B(v)}\left(\delta_k^+\Gamma_{k,k}^{++} + \delta_{k+1}^-\Gamma_{k,k+1}^{+-}\right),$$

$$\tag{6.51}$$

which replace the fallacious Eqs. 6.42 and 6.45. With the aid of Eq. 6.49, this system reduces to

$$n\frac{B(v) - B}{B(v)}\left(\left(\Gamma_{k+1,k}^{-+} - \Gamma_{k,k}^{++}\right)\delta_k^+\right.$$

$$\left. + \left(\Gamma_{k+1,k+1}^{--} - \Gamma_{k,k+1}^{+-}\right)\delta_{k+1}^-\right) = -D(v). \tag{6.52}$$

The solution of the system (Eqs. 6.52 and 6.50) has the form

$$\delta_{k+1}^-(B) = \frac{1}{\gamma(k)}\left(\gamma_k^+ - \frac{B(v)}{B(v) - B}D(v)\right),$$

$$\delta_k^+(B) = \frac{1}{\gamma(k)}\left(\gamma_{k+1}^- + \frac{B(v)}{B(v) - B}D(v)\right), \tag{6.53}$$

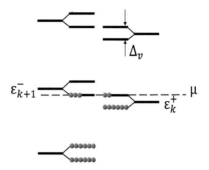

Figure 6.8 Schematic diagram of merging of the spin- and valley-split Landau levels at the chemical potential in MOSFETs. The occupied levels are indicated by red dots. Reprinted (figure) with permission from [Shashkin, A. A., Dolgopolov, V. T., Clark, J.W., et al. (2014). *Phys. Rev. Lett.* **112**, 18, p. 186402.] Copyright (2014) by the American Physical Society.

where $\gamma(k) = \gamma_k^+ + \gamma_{k+1}^- \equiv n(\Gamma_{k,k}^{++} + \Gamma_{k+1,k+1}^{--} - \Gamma_{k+1,k}^{-+} - \Gamma_{k,k+1}^{+-})$. In the region of the phase diagram where the level ordering is reversed, that is, $\epsilon_{k+1}^-(v) - \epsilon_k^+(v) < 0$, manipulations analogous to those above lead again to Eq. 6.53.

As mentioned above, standard FL theory says that at $B = B(N+1)$, one of the relevant Landau levels should be fully occupied, while its counterpart, completely empty, generates the jump of the magnetic moment M. By contrast, according to Eq. 6.49, at $B = B(N+1)$, energies of both the levels coincide, that is,

$$\omega_c(N+1) - \Delta_Z = 0. \tag{6.54}$$

and therefore the magnetic moment M has no jump at this point. Completion of the filling of one of the two levels occurs at $B < B(N+1)$. At this point, the levels are still coincident in energy, so, according to Eq. 6.38, no jump of M occurs at this point. As the density changes, quasiparticles (quasiholes) occupy the second, unfilled quantum level. Thereby in the whole interval $B(N+2) \leq B < B(N)$, there exists just *one* jump of M, which occurs at $B = B(N+2)$ where the filling of both levels is completed. Thus we infer that the onset of merging doubles the oscillation period. We notice that when the merging of the two levels is ended, all other crossings occur *between filled or empty levels*, so the straight line $v = $ const. is ended at the point where Eq. 6.54 is met.

In dealing with electron systems of MOSFETs, which are specified by the presence of two valleys in the electron spectrum (see Fig. 6.8, Shashkin et al., 2014), analogous arguments lead to the following relation

$$\omega_c(N+1) - \Delta_Z - \Delta_v = 0 \qquad (6.55)$$

(with Δ_v, being the valley splitting) for the boundary $n_m(B)$ of the region where the oscillations disappear. Quantities ω_c, Δ_Z and Δ_v are known from independent experiments. The analysis of this equation performed by Shashkin et al. (2014) is presented in Fig. 6.9. Theoretical results for the upper boundary of the merging region $n_m(B)$, shown by the solid blue line, are seen to be in agreement with the experimental density $n^*(B)$ at which the oscillation minima at $v = 4k + 4$ disappear.

We now discuss the impact of the violation of the equidistance of electron spectra in strong magnetic fields on the structure of magnetic oscillations near the critical density n_c. As discussed in Section 6.3, close to the Fermi surface, a leading NFL term in the electron spectrum is the cubic corrections $\propto (p - p_F)^3$ that leads to the decline the effective mass m^* with the magnetic field, which is important near the topological transition point where the linear term $\propto (p - p_F)$ is strongly suppressed to yield

$$\epsilon(p) = \frac{p_F(p - p_F)}{m^*} + \beta \frac{(p - p_F)^3}{3m_b p_F}, \qquad (6.56)$$

where $\beta > 0$ is a coefficient. In external magnetic fields, the cubic correction leads to an additional term in the single-particle Hamiltonian of the form

$$\mathcal{H} = \beta \frac{[(\mathbf{p} - e\mathbf{A}/c)^2 - p_F]^3}{24 m_b p_F^4}. \qquad (6.57)$$

where $\mathbf{A}$ is the vector potential. The resulting correction to the spectrum leads to a modification of Eq. 6.55 that describes the upper boundary of the merging region. The corrected equation reads (Shashkin et al., 2014)

$$\omega_c - \Delta_Z - \Delta_v = -\beta \frac{(eB)^3}{3m_b c^3 (\pi n_m)^2}. \qquad (6.58)$$

Apparently, the right-hand side of Eq. 6.58 can be important at high magnetic fields. Taking in agreement with results of Section 6.3, β

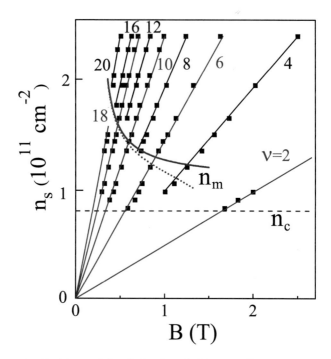

Figure 6.9 Positions of the Shubnikov–de Haas oscillation minima in the (B, n) plane (squares) and the expected positions of the cyclotron and spin minima calculated according to the formula $n = veB/c$ (solid lines). The position of the metal–insulator transition in $B = 0$ is indicated. The calculated merging boundary $n_m(B)$ is shown by the solid blue line for $\beta = 0$ and the dotted violet line for $\beta = 1$ in Eq. 6.58. Reprinted (figure) with permission from [Shashkin, A. A., Dolgopolov, V. T., Clark, J.W., et al. (2014). *Phys. Rev. Lett.* **112**, 18, p. 186402.] Copyright (2014) by the American Physical Society.

$= 1$, we determine the corrected dependence $n_m(B)$, shown by the dotted violet line in Fig. 6.9. As seen, in dealing with strong magnetic fields accounting for cubic corrections to the electron spectrum leads to better agreement between theoretical and experimental results.

As a matter of fact, doubling of the oscillation periods in MOSFETs is not always the ultimate step of the rearrangement of the oscillation pattern near the critical density n_c, where the Landau state loses the topological stability. Sometimes the presence of two almost degenerate valleys in the electron spectrum results in the complication of this pattern due to the occurrence of the fine structure of the rearranged magnetic oscillations.

To analyze this fine structure, we note that each episode of merging of the levels ϵ_{k+1}^- and ϵ_k^+ is ended when the distance between these levels becomes larger than the matrix element of the interaction amplitude $\gamma(k)$ and consider the case $\Delta_v > \gamma(k)$. For the sake of simplicity, we restrict ourselves to the case of strong magnetic fields, considering the lowest values of $v \leq 4$. In what following we employ notations 0^{--}, 0^{-+}, 0^{+-}, and 0^{++}, where the first upper index refers to the spin projection of the level, while the second is its valley index, the lowest level being 0^{--}, and 0^{++}, the highest one. To begin let us address the case $v = 2$ of two quantum levels 0^{--} and 0^{-+}, with the splitting Δ_v, whose value is assumed to be small compared with the cyclotron and spin splittings. This case is schematically illustrated in the top left panel of Fig. 6.10. In this case, the standard FL structure of Landau-level fan diagram with the continuous straight line $v = 2$ holds. However, in the case $v = 3$, the situation changes as the critical density n_c approaches and the distance between the lowest empty levels 1^{--} and the highest filled level 0^{+-} shrinks (see top right panel of Fig. 6.10). The analysis, analogous to that carried out above, shows that the straight line $v = 3$ is ended by virtue of merging these levels at a critical density, at which the distance between them vanishes. The oscillation $v = 3$ is recovered at a point where the levels 1^{--} and 0^{+-} become completely separated, the former being filled, while the latter is empty.

In the most interesting case, $v = 4$, where at high electron densities, all 0^{--}, 0^{-+}, 0^{+-}, and 0^{++} are completely filled, the first catastrophe of FL theory occurs, when the upper filled level 0^{++} crosses the nearest empty level 1^{--} that triggers their merging with the subsequent disappearance of the $v = 4$ magnetic oscillation (see columns a and b of the bottom panel of Fig. 6.10). In case the magnitude of the splitting Δ_v is large enough, there exists a critical point where the episode of merging is ended, and the level 1^{--} turns out to be completely filled, while the level 0^{++} is empty (this point is shown in column c of the bottom panel of Fig. 6.10). At this point, the $v = 4$ oscillation is recovered, as in the previous

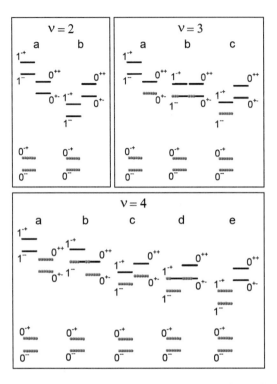

Figure 6.10 The scheme of six lower Landau levels for different ν and decreasing density n. Top left: $\nu = 2$. Two levels 0^{--}, 0^{-+} stay filled and the FL structure of the fan diagram holds. Top right: $\nu = 3$. At high density three levels 0^{--}, 0^{-+}, and 0^{+-} are filled (column a). As density decreases, merging of the levels 1^{--} and 0^{+-} occurs, and the oscillation disappears (column b). As density further drops, approaching n_c, the oscillation is recovered (column c). Bottom: $\nu = 4$. At high density, four levels 0^{--}, 0^{-+}, 0^{+-}, and 0^{++} are filled (column a). With decreasing density the oscillation disappears due to merging of the levels 1^{--}, 0^{++} (column b) and is recovered when the episode of merging is ended (column c). When n further decreases, the merging of the levels 1^{-+} and 0^{+-} occurs that results in the second disappearance of the oscillation (column d) and then ends (column e) with recovering of the straight fan line.

case, $\nu = 3$. Nothing awful then occurs in the case where two filled levels 1^{--} and 0^{+-} cross each other, as well as two empty levels 1^{-+} and 0^{++}. A new catastrophe takes place at a critical density, at which the empty level 1^{-+} merges with the filled level 0^{+-}. Again the $\nu = 4$ oscillation disappears (see column d of the bottom panel of Fig. 6.10). It is recovered at a point where the merging of the levels 1^{-+} and 0^{+-} is ended. In doing so the level 1^{-+} turns out to be completely filled, while the level 0^{+-} is empty (column e). Conditions that promote this NFL scenario, including requirements to the magnitude of corresponding matrix elements of the scattering amplitude, will be analyzed elsewhere.

6.5 Conclusion

We have developed a microscopic theory of a 2D electron gas and applied it to the elucidation of one of the brightest experimental discoveries of the last decade—the divergence of the electron effective mass m^* in a 2D homogeneous electron gas, uncovered in Si-MOSFETs where this divergence occurs at $r_s \approx 8 \div 9$. The theory is based on the functional approach (Khodel et al., 1994), which is ab initio, and nonperturbative where there is neither fitting nor smallness parameters. This renders the theory applicable at any value of the spacing parameter r_s below the critical value r_{sc}, at which the Landau state loses its topological stability. The efficiency of the theory is confirmed by a good agreement of our results for the static linear response function and correlation energy of the 2D electron gas with available results of MC simulations. Within the developed theory, we have also calculated the single-particle spectrum $\epsilon(p)$ of a homogeneous 2D electron gas and demonstrated that the dispersion of the spectrum $\epsilon(p)$ at the Fermi surface rapidly declines with increasing r_s, the effective mass diverging at $r_s^\infty \simeq 7$.

Flattening of the electron spectrum near the critical point where m^* diverges suggests an idea that near the TT point in heterostructures where damping of single-particle excitations is extremely small and electron–phonon attraction is present, a 2D electron gas should undergo a superconducting phase transition. The underlying reason for it is associated with huge enhancement of the density of states, triggered by flattening of the single-particle spectrum. With the aid of formulas obtained by Khodel et al. (1994, 2007), the estimated critical temperature is of order of 1 mK or even higher.

Acknowledgments

We are grateful to J. W. Clark, V. T. Dolgopolov, V. R. Shaginyan, and A. A. Shashkin for fruitful discussions. This work was partially supported by grant no. NS-932.2014.2 of the Russian Ministry for Science and Education and by RFBR grant no. 15-02-06261, as well as the McDonnell Center for the Space Sciences.

References

Abrahams, E., Kravchenko, S. V., and Sarachik, M. P. (2001). *Rev. Mod. Phys.* **73**, 2, pp. 251–266.

Abrikosov, A. A. (1988). *Fundamentals of the Theory of Metals* (North-Holland, Amsterdam).

Bäuerle, C., Bunkov, YU. M., Chen, A. S., Fisher, S. N., and Godfrin, H. (1998). *J. Low Temp.* **110**, 1/2, pp. 333–338.

Borisov, V. V., and Zverev, M. V. (2005). *JETP Lett.* **81**, 10, pp. 503–508.

Casey, A., Patel, H., Nyeki, J., Cowan, B. P., and Saunders, J. (2003). *Phys. Rev. Lett.* **90**, p. 115301.

Deviatov, E. V., Khrapai, V. S., Shashkin, A. A., et al. (2000). *JETP Lett.* **71**, 12, pp. 496–499.

Dolgopolov, V. T. (2007). *Low Temp. Phys.* **33**, 2, pp. 98–104.

Fang, F. F., and Stiles, P. J. (1968). *Phys. Rev.* **174**, pp. 823–827.

Gegenwart, P., Si, Q., and Steglich, F. (2008). *Nat. Phys.* **4**, 3, pp. 186–197.

Kapustin, A. A., Shashkin, A. A., Dolgopolov, V. T., et al. (2009). *Phys. Rev. B* **79**, p. 205314.

Kawazoe, Y., Yasuhara, H., and Watabe, M. (1977). *J. Phys. C: Solid St. Phys.* **10**, 17, pp. 3923–3304.

Khodel, V. A., and Shaginyan, V. R. (1990). *JETP Lett.* **51**, 9, pp. 553–555.

Khodel, V. A., Shaginyan, V. R., and Khodel, V. V. (1994). *Phys. Rep.* **249**, pp. 1–134.

Khodel, V. A., Clark, J. W., Li, H., and Zverev, M. V. (2007). *Phys. Rev. Lett.* **98**, 21, p. 216404.

Khodel, V. A., Clark, J. W., Popov, K. G., and Shaginyan, V. R., (2015). *Pis'ma v ZhETF* **101**, 6, pp. 448–454.

Kravchenko, S. V., Shashkin, A. A., Bloore, D. A., and Klapwijk, T. M. (2000). *Solid State Commun.* **116**, 9, pp. 495–499.

Kravchenko, S. V., and Sarachik, M. P. (2004). *Rep. Prog. Phys.* **67**, 1, pp. 1–44.

Kwon, Y., Ceperley, D. M., and Martin, R. M. (1993). *Phys. Rev. B* **48**, 16, pp. 12037–12046.

Landau, L. D. (1956). *Sov. Phys. JETP* **3**, 6, pp. 920–925.

Landau, L. D. (1959). *Sov. Phys. JETP* **8**, 1, pp. 70–74.

Lifshits, E. M., and Pitaevski, L. P. (1980). *Course of Theoretical Physics, Volume 9: Statistical Physics, Part 2* (Pergamon, New York).

Löhneysen, H., Rosch, A., Vojta, M., and Wölfle, P. (2007). *Rev. Mod. Phys.* **79**, 3, pp. 1015–1076.

Melnikov, M. Yu., Shashkin, A. A., Dolgopolov, V. T., Kravchenko, S. V., Huang, S. H., and Liu, C. W. (2014). *JETP Lett.* **100**, 2, pp. 114–119.

Migdal, A. B. (1967). *Theory of Finite Fermi Systems and Applications to Atomic Nuclei* (Wiley, New York).

Mokashi, A., Li, S., Wen, B., et al. (2012). *Phys. Rev. Lett.* **109**, 9, p. 096405.

Moroni S., Ceperley D. M., and Senatore G. (1992). *Phys. Rev Lett.* **69**, 13, pp. 1837–1840.

Neumann, M., Nyeki, J., Cowan, B. P., and Saunders, J. (2007). *Science* **317**, 5843, pp. 1356–1359.

Pines, D., and Noziéres, P. (1966). *Theory of Quantum Liquids* (Benjamin, New York).

Pudalov, V. M., Gershenson, M. E., Kojima, H., et al. (2002). *Phys. Rev. Lett.* **88**, p. 196404.

Shaginyan, V. R. (1985). *Solid State Commun.* **55**, 1, pp. 9–12.

Shashkin, A. A., Kravchenko, S. V., Dolgopolov, V. T., and Klapwijk, T. M. (2002). *Phys. Rev. B* **66**, p. 073303.

Shashkin, A. A. (2005). *Phys. Usp.* **48**, 2, pp. 129–150.

Shashkin, A. A., Dolgopolov, V. T., Clark, J. W., et al. (2014). *Phys. Rev. Lett.* **112**, 18, p. 186402.

Stern, F. (1967). *Phys. Rev. Lett.* **18**, 14, pp. 546–548.

Zhang, Y., Yakovenko, V. M., and Das Sarma, S. (2005). *Phys. Rev. B* **71**, 11, p. 115105.

Zverev M. V., Khodel, V. A., and Shaginyan, V. R. (1996). *JETP* **82**, 3, pp. 567–575.

Index

2D electron gas 135, 189–191,
    195–196, 199–201, 215
2D electron system 47–50, 56,
    60–61, 69–70, 72, 74, 76, 82,
    84, 98, 109, 145, 148, 168,
    173, 177, 203–205
^{3}He monolayers 1, 3, 16–18

activation energy 51–52, 150, 152,
    163
aging 160, 172, 174–175, 178, 182
analytic continuation 77–79
Anderson localization 6, 25, 49,
    65–68
Anderson transition 66–67, 158
antiferromagnetic 2, 73, 125–126,
    131, 140, 179
antiferromagnetic order 126
antisymmetric state 98–100, 108
approximations 7, 24, 68, 194,
    200–201, 208
    local 24, 194, 198, 202
    one-loop 74, 79–80, 105, 109
asymmetry 100, 108
atomic orbitals 37–38

band structures 123–124
bandwidth 24, 41, 60–61,
    122–123, 125
    electronic 19, 26, 117
behavior 28, 31, 34–35, 50, 53,
    58–61, 65, 70–74, 77, 81–82,

89, 94, 96–97, 110, 131,
    136–137, 146–148, 152,
    154–155, 157–159, 163,
    166–167, 169–170, 174–175,
    177–180, 190–191, 203
    complex 61, 173
    critical 16, 118, 146, 148–149,
        178, 182
    crossover 26, 34, 135
    glassy 148, 157, 160, 165, 172,
        179
    insulating 5, 8, 25–27, 93
    metallic 3, 69, 90, 94, 98, 147,
        154, 158
    QC 21–22, 37
    quantum-critical 8, 136–137
    resistance 89, 94
    scaling 5, 8, 37, 127, 146–149,
        157, 162
    spin liquid 17, 21
boundary 124, 127, 211–212
branches 8, 136, 150
    insulating 122–123, 135–138
    metallic 122–123, 135–138
Brillouin zone 72, 124
Brinkman–Rice scenario 120

calculation 10, 29, 37, 40–41, 55,
    66, 195–197, 202
    MC 195
    RPA 195–196, 200
carrier concentration 152, 181
catastrophe 213–214

channel
  Cooper 75, 82
  particle–hole 67
  particle–particle 67
  triplet 86, 101–102
charge 24, 29, 75, 162, 173,
  178–179, 191–192
charge carriers 161, 179–180
charge dynamics 145, 149, 157,
  165, 171–173, 177–179, 182
charge ordering 41
coefficient 59, 107, 211
coexistence dome 20, 34–36, 40
compounds 2, 20–21, 25, 127,
  131, 135
  heavy-fermion 6, 11, 29, 33,
  190
concentration 96, 106–107
conductance 66, 68–69, 130, 177,
  179
  average 158
  non-Gaussian 174–175
conductivity 48, 56, 60, 68–69, 77,
  81–84, 97, 105–107, 125,
  128–130, 146–151, 153, 155,
  163, 170–171, 173, 182
  critical 147–148, 151, 157
  Drude 75
  extrapolated
    zero-temperature 155
  temperature-dependent 57
  time-averaged 158–159, 163,
  176
conductivity exponent 147–148
conjugate field 127
constraints 74, 173, 175
contribution 69, 75, 81–82, 158,
  192–193, 198–199, 201–202,
  204
  weak localization 69, 81–83
cooling 175–176
cooperons 67, 74–75, 81–82
corrections 23, 75, 82–83, 97, 130,
  158, 195, 202, 211–212

correlation 195, 201
  cubic 211–212
  exchange 195, 201
  interaction 106–107
  localization 106, 108
correlation effects 6–7, 9, 19, 24,
  29
correlation length 134, 146, 148
Coulomb glasses 25, 145, 158,
  172, 175, 178, 182
Coulomb interactions 6–7, 20,
  22–23, 71, 86–87, 101,
  148–149, 157, 161–165, 167,
  177–178, 182
  electron–electron 162
  long-range 157, 162–163, 182
  screened 86, 162, 164
Coulomb repulsion 15, 18, 25–26,
  29, 34, 38–39, 120, 122, 155
critical conductivity 147–148,
  151, 157
critical density 5–6, 9, 11, 13,
  50–52, 57, 147, 150, 152, 155,
  159, 163, 190, 203, 211–214
critical exponents 21, 58, 123,
  127–128, 134–135, 138–139,
  147, 154, 156, 160, 162–164,
  167–169, 171, 178, 181–182
criticality 72, 123, 130, 134
critical phenomena 35–36, 66,
  120–121, 127, 131, 140
critical temperatures 120, 122,
  125, 127, 131, 215
crossover, metal–insulator 5, 26,
  125, 132
crossover behavior 34, 135
  metal–insulator 25
crossover temperatures 7, 24, 137
crystals 18–19, 169
  organic Mott 18, 21
  Wigner 11, 29, 38–41, 48, 57,
  162
CuO_2 planes 179–180
cuprates 145, 149, 178–183

cyclotron 60, 203–205, 212–213
cyclotron splitting 53, 60,
    204–205

de Haas–van Alphen
    oscillations 206
degrees of freedom
    charge 118, 165, 178
    isospin 65, 73, 81, 83, 109
    spin 12, 39, 65, 72–73, 81, 83,
        109, 165, 177
delocalized states 66–67
densities 5, 8, 11, 13, 14, 38, 49,
        57, 60–61, 151, 159, 214
    carrier 5, 146, 148–149, 173,
        176
    critical 5–6, 9, 11, 13, 50–52,
        57, 147, 150, 152, 155, 159,
        163, 190, 203, 211–214
    doublon 127–129
    electron 12–14, 48–60, 70, 100,
        150, 182, 189–190, 195,
        203–205
    thermodynamic 75
dependence
    density 9, 70
    mass 28
    temperature 5, 7–8, 24, 30–31,
        50, 56, 61, 65, 68, 70–74, 76,
        83–84, 93, 97–98, 105, 109,
        170, 173
dephasing rate 83, 105
dephasing time 83–84, 105–107
derivative criterion 50–51
devices 4–5, 149–150, 161–163,
        169–170, 181
    high-disorder 162–163, 167
    low-disorder 162, 167, 169
diffusion coefficient 75, 102
diffusive motion 67
diffusive regime 6–7, 10–11, 15,
        101
dimers 19, 124–125

disorder 3–4, 6, 9, 11, 56–57,
        145–146, 148–150, 152,
        154–155, 157, 160, 167–169,
        172, 178, 181–182
    common 98, 109
    low 167
    weak 3–4, 49, 66–67, 69, 154
    weaker 7, 11
divergence 16–17, 27–29, 59, 60,
        67, 84, 89, 172, 190–191, 202,
        215
DMFT, *see* dynamical mean-field
    theory
DMFT phase diagram 35
DMFT predictions 21, 31, 138, 140
domains 179, 208
doping 178–180
double quantum well 98, 101, 105,
double-quantum-well
    heterostructure 65, 73–74,
        98–100, 104–107, 109
dynamical mean-field theory
    (DMFT) 6, 21, 23–27, 29,
        31–37, 40–41, 60, 120–122,
        127–132, 138–140
dynamical scaling 150, 154, 157,
        160, 171, 173, 182

effective mass 14, 16–17, 19,
        23–24, 27, 39–40, 47–48, 53,
        55–60, 75, 190–191, 201–205,
        211, 215
effective mass divergence 28–29
effective mass enhancements 19,
        23, 27–31, 56
electron annihilation operator 84,
        98
electron backscattering 49
electron concentration 70–71, 73,
        84, 94, 98, 101–102, 105–108
    equal 73, 98, 101–102, 105, 108
electron correlation 118, 123

electron densities 12–14, 48–49,
    51–52, 54–60, 70, 100, 150,
    189–190, 195, 204
  critical 50–52, 56
  decreasing 58
  high 203, 205, 213
  low 49, 53, 56, 181, 205
electron density dependence 70
electron dynamics 6, 158
electron–electron interactions
    48–49, 56, 58, 68–69, 75,
    87–88, 91, 100–101, 146, 148,
    152, 154, 157, 160–161, 169,
    182
  strong 3, 49, 69
electron–electron scattering 68
  inelastic 6–7, 31
electron–electron scattering
    amplitude 207
electron gas 135, 189–191,
    194–196, 199–202, 215
  interacting 38, 190
  noninteracting 192, 204–205
electronic correlations 1, 4, 6, 125,
    149, 172
electronic phases 117, 123, 140
  correlated 169
electron–phonon interaction 68
electronic quantum-critical
    fluid 140
electronic states 4, 66, 140
  bound 15, 38
electronic systems 25, 41
  correlated 24, 33
electron liquid 1, 6, 48, 65, 74, 203
electron operators 76, 86, 99, 101
electrons 4, 7, 13, 25–26, 29,
    37–39, 47–49, 52, 58, 66–69,
    72–73, 75, 83–85, 87–88,
    90–91, 98–100, 102, 104–106,
    109, 117, 119–120, 127, 140,
    148, 150, 152, 155, 161–162,
    172, 175, 182, 190, 204
  conduction 30, 154

  correlated 24, 33, 37, 119–120,
    169
  interacting 38, 69, 83, 190
  mobile 2, 9
  noninteracting 4, 58, 66, 68,
    75–76
  spinless 71
electron spectrum 196, 211–213,
    215
electron spin 12, 55, 70–71
electron systems 49, 55, 65, 81,
    83, 101, 189, 190–191, 203,
    211
  disordered 49, 84, 97, 109
  low-disorder 49, 58
  single-valley 90
  two-valley 72, 74, 81, 85, 87,
    94, 109
energies 66, 72, 87, 122, 139–140,
    158, 177, 207, 210
  ground-state 190, 194
  Matsubara 74, 76
  on-site Coulomb 117, 119
  repulsive 117, 119
entropy 28–31, 121, 126
  residual 28–29
  spin 29–30, 126
evaluation 191, 194–198
exchange interactions 29
exchange processes 17, 29
excitations 6, 15, 24, 38–39, 126,
    172, 181
  single-particle 26, 39, 190, 196,
    215
experimental data 10, 72, 96, 106,
    108, 110, 190, 202–203
experimental discoveries 8, 215
experimental estimates 107–108
experimental evidence 145–146
experimental features 3, 65, 154
experimental investigations 12,
    37, 190
experimental observations 69, 94,
    131, 159, 167

experimental signatures 4, 148
experimental values 130, 139
experimental work 13, 21
experiments 1, 9–10, 13–14,
    28–29, 31–32, 35–36, 39–40,
    59, 73–74, 93–94, 96–99,
    107–109, 128, 150, 154, 161,
    172–173, 175, 178, 181–182,
    211
exponents 128, 171
  critical 21, 123, 127, 128, 139,
    147, 154, 160, 162–163,
    167–169, 171, 178, 181–182
extrapolation 50, 52, 154, 157,
    159, 163, 165

Fermi energy 18, 34, 38, 48,
    60–61, 149
Fermi level 52, 60–61, 204, 209
Fermi liquid (FL) 3, 7, 11, 16, 26,
    30, 37, 53, 60, 68, 75–76, 86,
    101, 119, 136–138, 140,
    158–159, 190
  ferromagnetic 48
  paramagnetic 48
Fermi liquid theory 58
Fermi surface 25, 101, 124–125,
    130, 199, 201–202, 204, 207,
    209, 211, 215
Fermi temperature 16, 20, 27, 32,
    34, 150
ferromagnetic transition 53
FL, *see* Fermi liquid
FL pattern, standard 206–207
FL picture 7, 11, 14, 24
fluctuations 30, 36, 139, 158,
    172–173, 175–177
  quantum 120–122, 134, 146
  quantum-critical 122, 134–135,
    139
  superconducting 180–181
  thermal 27, 121, 134, 147

free-energy landscape 172, 175,
    177
frequencies
  Matsubara 76
  real 77–78
frequency 83–84, 204–205

gate 106, 162, 165
g-factor 14, 17, 48, 53, 55–57, 84
glasses 172, 174, 179
  charge 158, 172
  Coulomb 25, 145, 158, 172,
    175, 178, 182
  spin 175
glassiness 157, 173–175, 179
glass transition temperature 173
glassy behavior 148, 157, 160,
    165, 172, 179
glassy dynamics 172, 175,
    177–179
glassy systems 172, 174, 177
graphite 16–17
ground states 12, 21, 25–26,
    48–49, 119, 121, 131, 135,
    137, 146–147, 150, 161, 163,
    178–180, 203
ground-state energy 190, 191,
    194, 204

half-filled bands 117, 123–125
Hamiltonian 22, 24, 67, 84, 87, 98,
    146, 191
  single-particle 67, 85, 99, 211
heterostructures 106–107, 162,
    189, 215
  double-quantum-well 65, 74,
    99
  semiconductor 149, 169, 179,
    181
hierarchical pictures 175, 177,
    179

high-disorder samples 157–158,
161, 163–165, 167, 169, 171,
177, 182
Hubbard model 17, 22–23, 37, 40,
60, 119, 138
half-filled 18, 26, 32, 35, 37
single-band 22, 37

inelastic electron–electron
scattering 6–7, 31
instability 17, 55, 199, 200, 202
insulating branches 122–123,
135–138
insulating phase 14, 18–19, 26, 29,
50, 119–120, 125–126, 128,
149–150, 155, 171, 174
insulating state 8, 13, 25, 29, 38,
40–41, 118, 171–172, 182
insulators 2
interaction amplitudes 75–76, 81,
83–84, 87–89, 92, 97,
103–105, 107, 109, 213
interaction corrections 106–107
interaction energy 192–193
interaction parameters 48–49, 86,
100–102, 107
interaction, effective 192–194,
196–198, 200–201
interface 149, 155, 189
intermediate length scales 88–91
intermediate phase 40, 48, 61,
160, 166, 182
intermediate-temperature
range 35–36, 122, 131
intermediate-temperature
region 26, 35, 135
invariance 79, 88, 91, 104, 174
ionic liquid 181
ionic potential 38
ions 37–38, 149, 191
irregularities 199–201
isospin indices 74–75
isospin projection 85, 91, 109

isospin space 75, 80, 86, 100

jump 17, 204, 206–207, 210
jump magnitude 204

kinetic energy 26, 39, 48, 117, 119

Landau levels 191, 203–205, 210,
214
filled 206–207
Landau state 191, 199, 202–203,
207, 212, 215
lattice, crystal 38
lattice constant 85
lattice geometry 123, 139
lattice site 16, 26, 29, 119, 127
length scale 67, 76, 79, 81–85,
87–92, 97, 100, 104, 106, 110,
134, 161, 165, 179
linear response 77–78, 91
linear response function 191–193,
198, 215
localization 9, 49, 67, 69, 181
Anderson 6, 25, 49, 65–68
Mott 22
strong 48, 150
weak 48–49, 68–69, 81–83,
106, 108
localization length 50, 52, 67
temperature-independent 53
localization, scaling theory of 4,
48, 161
local magnetic moment 26, 29–30,
151, 153–157, 162, 165–171
local moments 26, 31, 39, 155,
165, 167, 171
low-disorder samples 52–53,
150–152, 154, 156, 160, 163,
167, 171, 178, 182
low-temperature behavior 72, 136

low-temperature
    magnetoresistance 12, 54
low-temperature resistivity 11, 65

magnetic moment 13–15, 23, 26,
    29–30, 38, 41, 151, 153–158,
    162, 165–171, 179, 204,
    206–207, 210
magnetic field 11–12, 27, 53–55,
    71, 82, 96, 108, 126, 146, 165,
    179–180, 191, 205, 211
    perpendicular 11, 57, 67, 75,
        82–82, 203
    strong 16–17, 211–213
    parallel 11–13, 40–41, 54–55,
        57, 71, 73, 83–85, 94–97, 109,
        163, 166
    weak 204, 207–208
    zero 3, 13, 39–40, 48–50,
        52–53, 56
magnetic frustration 19–20
magnetic order 15, 17, 19–20,
    22–23, 126–127
magnetic ordering 15, 25–27, 34
magnetic oscillation 191, 205,
    207, 211, 213
magnetization measurement 57
magnetoconductance 152
magnetoresistance 53, 72, 83, 96,
    106, 108, 117, 180
    normalized 53–54
    parallel-field 53, 57
    positive 27, 53
mass
    band 32, 38, 48, 59, 201–202
    bandwidth-related 60–61
    effective 14, 16–17, 19, 23–24,
        27–31, 38, 40–41, 47–49, 53,
        55–61, 75, 190–191, 201–205,
        211, 215
matrix element 74, 85, 99,
    207–208, 213, 215
MC, *see* Monte Carlo

mechanism
    physical 3, 7, 9
    disorder-screening 10–11
    scattering 33
mean free path 66–67, 159
melting 39, 145, 162, 178
    quantum 38–39
    thermal 39
    Wigner crystal 40
metal–insulator transition 1–3,
    38, 47, 49–50, 53, 55, 57–58,
    69–73, 84, 117–118, 135,
    139–140, 145, 167, 169, 171,
    190, 205, 212
    intermediate 2
    quantum 139
    two-dimensional 3
metallic behavior 69, 90, 94, 98,
    147, 154, 158
    non-Fermi liquid 3
metallic branches 122, 136–137
metallic curves 10
metallic gate 161, 162
metallic glassy phase 160, 163,
    165–166, 172, 174, 177, 182
metallic ground state 49, 150, 161
metallic phase 8–9, 12, 41,
    119–120, 125–128, 147, 154,
    159–160, 165–167, 182
metallic regime 11, 32, 151–152,
    154, 162, 171
metallic region 72–73
metallic state 50, 138, 150
metal-oxide-semiconductor
    field-effect transistor
    (MOSFET) 49, 70, 135, 149,
    162, 189–191, 201, 203,
    205–207, 210, 211–212
metals 2, 6–7, 13, 25, 154, 160,
    165, 182
    correlated 3, 27
    spin-polarized 166
metastable states 172, 175

mirror symmetry 5, 8–9, 21,
     36–37, 99, 135, 137–138
mobilities 98, 101–102, 105, 108,
     169
mobility edge 51–52, 66
models 4, 18, 22, 25, 41, 148, 169,
     174
    Hubbard 17–18, 22–23, 26, 32,
     35, 37, 40, 60, 119, 138
    jellium 191–192
    minimal 119
    physical 189
molecular arrangements 123–125
molecular packing 123, 125
momenta 199, 202
momentum 83–84, 87, 101, 192,
     196, 199, 208
    Fermi 66, 87, 101, 190
monolayers 17
^{3}He 1, 3, 16–18
Monte Carlo (MC) 31, 39, 60,
     190–191, 195
MOSFET, *see* metal-oxide-
    semiconductor field-effect
    transistor
Mott insulator 15–16, 20, 23, 29,
     37, 38, 41, 121, 126–127, 140,
     178
    gapped 137–138
Mott insulator-to-metal
    crossovers 132
Mott organics 18, 32, 35–36, 41
Mott systems 19, 23, 32–34, 37,
     41
    organic 9, 18, 29, 33, 37, 131
Mott transition 1, 11, 15–19,
     21–22, 24, 26–30, 32–38, 41,
     117–123, 125–132, 135, 140
    first-order 120, 125–126, 133
    quantum 139

nearest-neighbor sites 119, 139
NFL, *see* non-Fermi liquid

noise 171–172, 174–176, 179
non-Fermi liquid (NFL) 3, 7, 158,
     160, 163, 165, 190
noninteracting electrons 4, 58, 66,
     68, 75–76
noninteracting electron gas 192,
     204–205
nonlinear σ-model 66, 74
nonmagnetic spin liquid behavior
     17

occupation 38, 209
    double 38, 117–118
one-loop renormalization 78–79
one-loop RG equation 81–82, 87,
     89, 103–104, 109–110
orbitals 18, 118
    atomic 37–38
    localized 38
order parameter 34, 118, 127,
     129–130
organic charge transfer salts 1, 3,
     18, 120, 123
organic materials 21, 123, 131,
     139
oscillation periods 203, 210, 212
out-of-equilibrium systems 172,
     174–175
oxides 27
    AFM 25
    copper 178
    transition metal 6

parallel magnetic field 11–13,
     40–41, 54–55, 71, 73, 83–85,
     94–97, 109, 163, 166
parameter 32, 34, 58, 66, 75,
     77–78, 83, 107, 118, 123, 134,
     139, 146, 148–149, 170, 181,
     197, 199–201, 215
    interaction 48–49, 76, 86–87,
     100–102, 107

order 34, 118, 127, 129–130
  scaling 54–55, 156, 160
Pauli spin susceptibility 57
perturbation 69, 105, 172–173, 175
perturbation theory 78
phase-breaking rate 68
phase-breaking time 68
phase diagram 13, 19–20, 26, 35–36, 40, 118, 120–122, 125–126, 132–133, 137–139, 147, 160, 165–166, 180, 206–207, 210
  experimental 40, 160
  temperature correlation 120
  temperature–pressure 125–126, 132–133, 137
phases 2, 11, 40, 48, 73, 119, 121, 136, 140, 147, 160, 182
  electronic 117, 123, 140, 169
  electronic Griffiths 7
  high-temperature 68, 121
  insulating 14, 18–19, 26, 29, 50, 119–120, 125–126, 128, 149–150, 155, 171, 174
  low-temperature 68, 121
  metallic 8–9, 12, 41, 119–120, 125–128, 147, 154, 159–160, 165–167, 182
  quantum spin liquid 126
phase transition 121, 146, 179, 209, 215
  continuous 6, 120, 146, 174
  discontinuous 120
  metal–insulator quantum 69
  quantum 8, 66, 121–122, 146
  thermal 146
physical content 8
physical mechanism 3, 7, 9
physical nature 2, 8, 34
physical observable 77, 79–82, 93
physical origin, common 18
physical processes 6, 25
  basic 22

physical reason 3
physical situation 38
physical systems 25, 27
physics 11, 23, 107, 119, 123, 175, 180
  condensed matter 145, 190
power law 66, 158, 160, 163
pressure 125, 127–128, 131–133, 146
  ambient 16
  critical 130
  external 125
  hydrostatic 16–17, 19, 123
  physical 123
  uniaxial 123
pressure dependence 131–132
projection 90, 92, 110, 208
  isospin 85, 91, 109
  spin 84–85, 88, 90–91, 98, 109, 213
propagator 78–80
  two-particle 67
protocol 149, 173–174, 176–177

QC, *see* quantum-critical
QC behavior 8, 21–22, 37
QCP, *see* quantum-critical point
QP, *see* quantum phase
QPT, *see* quantum phase transition
QSL, *see* quantum spin liquid
quantities 16, 32, 68, 75, 77, 101, 107–110, 165, 211
  physical 123, 125, 128, 134
quantum well 65, 72–74, 98–99, 102, 105–109, 190
  double 98, 101, 105
  single 101–102
quantum-critical (QC) 8, 22, 35, 121–123, 134–140, 147–148, 157, 162, 169, 171
quantum-critical behavior 136–137

quantum-critical point (QCP) 30,
34–36, 147–148, 157, 162,
171
quantum-critical regime 134–135
quantum-critical region 36, 123,
137–139, 147
quantum-critical scaling 22, 134,
139
quantum fluctuations 120–122,
134, 146
quantum levels 203–205, 207,
210, 213
quantum melting 38–39
quantum phase transition (QPT)
8, 15, 35, 66, 69, 121–122,
145–149, 155, 157, 159–160,
163, 165, 171, 182
quantum spin liquid (QSL)
126–127, 131, 140
quasiparticles 6, 14, 16, 27–28, 33,
85, 119–120, 196, 204,
207–210
heavy 16, 31, 33

random phase approximation
(RPA) 31, 86–87, 101–102,
190, 195–196, 200–201
rearrangement 172, 190–191,
203, 212
reciprocal space 72, 85
regime 2, 4, 6–8, 10–11, 14–15,
20, 24, 31, 48, 54, 89, 130,
150, 162, 179–180
correlated 31, 190
diffusive 6–7, 10–11, 15, 101
incoherent transport 6, 25
insulating 150, 157–158,
171–172, 174
low-temperature 20, 58,
137–138
metallic 11, 32, 151–152, 154,
162, 171
physical 6

QC 8, 21, 37
quantum-critical 134–135
quantum Hall 11, 39
quasiparticle 33
region 2, 55, 93, 122, 137, 140,
158, 180, 199, 202, 207,
210–211
coexistence 19–20, 26, 33–35
critical 8, 18, 59, 72, 146–147,
171
high-temperature 21
intermediate-temperature 26,
35, 37, 135
metal–insulator coexistence
33–34
QC 8, 36
quantum-critical 123, 137–139,
147
transition 23, 158
relaxation 84, 129, 172–176, 178,
182
renormalization 14, 79–80, 88,
202, 204
renormalization group (RG) 49,
60, 66
resistance 50–51, 72, 84, 88–90,
93–94, 96–98, 105, 179–180
resistance behavior 89, 94
resistivity 5, 7–12, 21, 27, 31–32,
35, 48–51, 53, 58, 61, 65,
69–74, 91, 95–97, 109,
131–132, 136, 203
critical 122, 181
normalized 133–137, 139
resistivity curve 5, 8, 20, 22, 36,
122
resistivity maxima 9–11, 20,
31–34
resistivity ratio 5
response function 190–195
linear 190, 192–193, 198,
215
RG, *see* renormalization group

RG equations 74, 81–82, 84, 88, 90, 92–93, 96–98, 104–105, 107, 109
  one-loop 81–82, 87, 89, 103–104, 109–110
RG flow 75, 89–90, 92–93, 97, 104–105
RPA, *see* random phase approximation
RPA calculation 195–196, 200
RPA value 201

samples 53, 57, 59, 107, 152, 156, 158, 162–163, 165, 169, 181
  high-disorder 157–158, 161, 163–165, 167, 169, 171, 177, 182
  low-disorder 52–53, 150–152, 154, 156, 160, 163, 167, 171, 178, 182
saturation 53, 136, 152
scaling 7–10, 13, 31, 32, 35–37, 54, 69, 128–129, 134–136, 148, 150–152, 156–158, 161, 163–164, 171, 173–174, 181–182
  quantum-critical 22, 134, 139
scaling analysis 21, 36, 122, 130–132, 134–135, 157, 163
scaling behavior 5, 8, 36–37, 127, 146–149, 157, 162
scaling curves 5, 135–138
scaling parameter 54–55, 156, 160
scaling relation 122–123
scaling theory 4, 21, 48, 66–67, 69, 71–72, 161
scaling function 8, 21, 32, 36, 122, 135, 138, 148
scatterers, common 73–74
scattering 6–7, 10, 24, 31, 33, 68, 72, 74, 84–86, 96, 98, 100, 108, 139, 149, 151, 153–157, 167, 170, 203
  intervalley 72, 84–86, 96
scattering amplitude 207–208, 215
scattering electrons 68, 85, 99
screening 7, 162–163
semiconductors 1, 3–4, 12, 22, 37, 48, 189
separation, phase 34, 40, 178
separatrix 50, 53, 150–153, 159, 166
Shubnikov–de Haas oscillation 60, 203, 212
silicon 2, 4–6, 10, 13, 32, 50, 54, 60, 145, 191, 201
silicon MOSFET 49–50, 53–54, 190
  low-disordered 12, 47, 51–52, 56, 60
Si-MOSFET 51, 65, 70–74, 76, 84, 87, 94, 96–97, 100–101, 109, 139, 149–152, 155, 158, 160, 162, 167–170, 172, 190, 202, 215
single-particle levels 204, 207, 209
single-particle spectrum 191, 196, 215
single quantum well 101–102
single-valley system 74, 90, 98
SIT, *see* superconductor–insulator transition
Sommerfeld coefficient 17, 28
specific heat 16, 28, 30, 68, 76–77
spectra, electronic 2, 24
spectrum, electron 196, 211–213, 215
spin polarization 41, 53, 55–57
spin projection 84–85, 88, 90–91, 98, 109, 213
spin-rotational symmetries 67
spin splitting 53, 94–96, 98, 108, 205–206, 213

spin susceptibility 13–14, 16–17,
40–41, 47, 53, 55–58, 60, 68,
88, 129
stability, topological 202, 207,
212, 215
stress 73
stretched-exponential form 8, 21
subband 155
substrate 153, 155, 158, 170
superconductivity 18, 117, 127,
178, 180
superconductor 23, 126, 131, 140,
179, 181
superconductor–insulator
transition (SIT) 181–182
symmetry 2, 67, 69, 199, 203
symmetry breaking 22, 34, 76–79,
81
symmetry class 67, 69
orthogonal 69

temperature behavior 70, 82, 96
temperature dependence 5, 8, 24,
30–31, 50, 56, 61, 65, 68,
70–74, 83–84, 93, 97–98,
105, 109, 170, 173
strong 8, 31, 68
weak 5
temperature range 5, 11, 20, 24,
148, 150
temperature 137–140, 146, 148,
150–151, 170–171, 173,
179–180
coherence 27–28, 30–31, 33
decreasing 49, 150, 154
finite 34, 48, 68, 82, 134
high 5–6, 20, 24, 87, 96,
107–108, 118, 148
low 7, 10, 13, 15–16, 26, 29, 54,
65, 68–69, 72, 74, 94, 107–
109, 118, 125, 131, 152, 181
lowest 24, 133, 151
zero 48, 50, 53, 68, 98

theoretical estimates 107
theoretical investigation 23, 119
thermal destruction 31, 33
thermodynamic response 13
thermopower 14, 47, 53, 58–61
TMO, *see* transition metal oxide
topological terms 67
top panel 155, 200–202
trajectory 35
transfer integrals 119, 123, 125
transition 6, 8–9, 11–13, 20, 22,
31, 33–34, 37, 39, 98–99, 120,
145–147, 150, 152, 171–172
finite-temperature 68
gas–liquid 118
glass 160, 165, 177
second-order phase 34, 146
transition metal dichalcogenides
169
few-layer 149
transition metal oxide (TMO) 6,
18, 22–23, 25, 33, 35, 37
transport 6–8, 13, 21, 31, 65, 72,
74, 158
low-temperature 72, 74, 109
quantum-critical 138, 140
transport experiments 74
transport measurements 56–57,
59
transport properties 18, 41, 66,
152, 155
two-valley electron system 72, 74,
81, 85, 87, 89, 94, 109

ultraclean limit 4
unit cell 25, 124
universality classes 12, 69, 128,
130, 139, 145, 147, 149, 157,
160, 167, 169, 182

vacancy interstitial pair 39

valleys 49, 72–73, 85, 108, 175,
    211, 213
valley splitting 72–74, 76, 84–85,
    91, 94, 96–97, 211
variations 96, 109, 129, 139, 192,
    197, 204, 207
voltage, threshold 52

weak localization correction
    108
weakly first order (WFO) 35
Wegner scaling 148, 151–152
WFO, *see* weakly first order

Wigner crystal 11, 29, 38–39, 40,
    48, 57, 162
Wigner crystallization 37, 41, 48,
    60–61, 135
Wigner lattice 38–39
Wigner–Mott physics 1
Wigner–Mott transition 1, 25,
    40–41

Zeeman splitting 12, 71, 76, 84,
    88, 93–94, 97
zero magnetic field 3, 13, 39–40,
    48–50, 52–53, 56